CW01665801

TEACH YOURSELF BOOKS

FURTHER CALCULUS

AND AN INTRODUCTION TO ANALYSIS

The object of this book is to introduce those who have a working knowledge of elementary calculus to wider and deeper aspects of the subject. It begins by clarifying such basic concepts as limit, continuity and infinity, and goes on to revise briefly differential calculus and integration. There then follow chapters on infinite series and integrals, the second mean value and other theorems, and differential and partial differential equations. It covers all the calculus required by students preparing for GCE 'A' level Mathematics.

TEACH YOURSELF BOOKS

FURTHER CALCULUS

AND AN INTRODUCTION TO ANALYSIS

F. L. WESTWATER
O.B.E., M.A.(Edin.), M.A.(Cantab.), A.M.I.E.E., R.N.(Rtd.)

TEACH YOURSELF BOOKS
ST. PAUL'S HOUSE WARWICK LANE LONDON EC4

First printed 1966
Second impression (with corrections) 1972

Copyright © 1966
The English Universities Press Ltd

All rights reserved. No part of this publication may be reproduced or transmitted in any form or by any means, electronic or mechanical, including photocopy, recording, or any information storage and retrieval system, without permission in writing from the publisher.

ISBN 0 340 05590 1

PRINTED AND BOUND IN ENGLAND
FOR THE ENGLISH UNIVERSITIES PRESS LTD
BY HAZELL WATSON AND VINEY LTD, AYLESBURY

PREFACE

The student who has studied elementary calculus may well wonder what comes next. He will expect to move on to harder examples, be introduced to partial differentiation, and struggle with multiple integration. These last two subjects really extend the ideas of the calculus which have so far been concerned with a single dimension or variable to include cases of two or more variables.

Even in the case of a single variable, progress is severely limited if we continue to think in finite terms. We have to learn to accept the idea of functions which can only be represented by an infinitely large sum of separate terms. We must also learn to extend the limits of integrals to include infinitely large values of the variable.

In the manipulative sense, it is true that these ideas do not introduce any essential difficulties. For example, partial differentiation is carried out in much the same way as ordinary differentiation and multiple integrals are formally evaluated by repeated integration. There is no particular difficulty in formally evaluating infinite limits of integration. It is, however, a fact that although from a practical point of view this may seem easy enough, these new concepts are really of far greater complexity and subtlety than those encountered in the simple case of one variable, and unless certain conditions are satisfied the formal procedures can lead to quite erroneous conclusions. It is, therefore, necessary to prove the validity of applying these processes and this cannot be done unless we clarify our concepts. It is for this reason that a great deal of time is spent in the first chapter of this book in an attempt to introduce precise notions of such concepts as infinity, continuity, and limits. Let it be said at once that it would be very difficult indeed to write on these matters without being strongly influenced by the work of G. H. Hardy, whose *Pure Mathematics* must remain a kind of bible in these matters. I gladly acknowledge the help I have obtained from this work, but, as the author himself states in the preface, it is essentially a book for mathematicians, and I have attempted to present some of the ideas in a simpler way, albeit at greater length.

The next two chapters deal with a revision of differential calculus and an extension to partial differentiation. This is followed by a revision of integration. I have tried to present this subject in a broader manner than is customary. In the early stages much attention is rightly paid to the acquisition of manipulative skill, but this can be overdone, and I have tried to show how it is possible in a few lines to show that a function is integable and even to see how it can be evaluated whereas the detailed steps may require two or three pages of foolscap and abound in possible sources of error.

The next chapter discusses in considerable detail the properties of infinite series and infinite integrals, and is followed by some additional theorems based on differential and integral calculus.

The book ends with two chapters on differential equations. The first of these deals with ordinary differential equations and covers the normal procedures employed in solving the more common types. It is beyond the scope of a book such as this to discuss partial differential equations in any detail and I have, therefore, confined myself to discussing their physical implications although simple solutions have been given in one or two cases.

F. L. W.

CONTENTS

CHAPTER ONE

BASIC CONCEPTS

Introduction

In dealing with the subject of the differential and integral calculus in an elementary manner, it is customary and entirely reasonable to avoid worrying the student with the problem of the precise meaning to be attached to such ideas as infinity, limit, continuous function, etc.

Most people have some intuitive understanding of these concepts and this serves them well enough when we are concerned only with an elementary treatment of the subject.

When, however, we wish to pursue the subject to a more advanced level it is essential to clarify our ideas and to attach precise meaning to the concepts of limit, continuity, infinity, and so on. This we shall now do.

To someone unfamiliar with more advanced mathematics, much of what follows may seem pedantic and some of the examples may appear highly artificial. From a certain point of view this may be true, but it is essential to bring home to the student as rapidly as possible the care necessary in defining our concepts. Examples which are not in the least artificial are easily found, but they are not always easy to understand. Surprisingly enough, intuition and common sense may be a good enough guide for most purposes but they can lead us badly astray in certain cases.

Before proceeding to develop our ideas in a logical manner we shall give an example of what we mean.

Consider the sum S_n in terms of the series

$$S_n = 1-\tfrac{1}{2}+\tfrac{1}{3}-\tfrac{1}{4}+\tfrac{1}{5}- \dots +(-1)^{n-1}1/n$$

We shall show in Chapter Six that if we imagine the number of terms n to increase indefinitely, the sum S_n can be made to differ by as little as we please from a fixed quantity S. i.e. For a very large value of n

$$S_n \text{ is approximately equal to } S$$

We can write this as

$$S = 1-\tfrac{1}{2}+\tfrac{1}{3}-\tfrac{1}{4}+\tfrac{1}{5}+\ \ldots \tag{1.1}$$

where the series is unending or, to put it popularly, 'goes on for ever'.

We shall show later that

$$S = \log_e 2 \quad \text{where} \quad e \text{ is the Naperian base.}$$

It must seem surprising to common sense when we state that the sum

$$1+\tfrac{1}{3}-\tfrac{1}{2}+\tfrac{1}{5}+\tfrac{1}{7}-\tfrac{1}{4}+\ \ldots = \tfrac{3}{2}\log 2 \tag{1.2}$$

when all we have done is to rearrange the terms of (1.1) so that two positive terms are followed by one negative term.

Or again, we can show that

$$1-\tfrac{1}{2}-\tfrac{1}{4}+\tfrac{1}{3}-\tfrac{1}{6}-\tfrac{1}{8}+\tfrac{1}{5}-\ \ldots = \tfrac{1}{2}\log 2 \tag{1.3}$$

where this arrangement has two negative terms following one positive term.

We hasten to add that where the number of terms in the series (1.1) is finite, i.e. terminates at some point, then rearrangement cannot alter the sum. Nonetheless when the number of terms is allowed to become indefinitely large, then the results are true.

We hope this example will convince the reader that great care is required in dealing with numbers which are allowed to become indefinitely large or which are popularly described as infinitely large. If such ideas are to lead to useful results then we must be sure of our foundations before developing any logical argument which involves them.

The rest of this chapter may seem a little heavy and dull,

full of definitions and proofs of statements which may appear obvious. Nonetheless it will repay careful study because it lays the foundation of what follows.

The Meaning of a Continuous Real Variable

Consider a real number x which we restrict to being between 0 and 1 by the relation

$$0 \leqq x \leqq 1 \tag{1.4}$$

Note that x may take the end values 0 and 1 as well as the values in the interval.

It will be obvious that any proper fraction of the form

$$x = \frac{m}{n} \quad \text{where} \quad m < n$$

satisfies (1.4) and that the number of such fractions is indefinitely large, i.e. we can never come to the end of such values. For let p and q be two fractions then $(p+q)/2$ is another lying between p and q and we can continue this process indefinitely.

In geometrical terms, if we think of each fraction as representing a point on the line joining the origin to the point with coordinate $+1$ on the x axis, then such points can be made indefinitely numerous and as close together as we please. If we could plot them then even through a powerful microscope we could make the line appear as a continuous one and not as a series of discrete points.

Nonetheless there are points on the line which do not correspond to fractions, and they too are indefinitely numerous. For example, the number $\frac{1}{2}\sqrt{2}$ cannot be expressed as a rational fraction, but it lies between 0 and 1 and is therefore one of the values which x can take. The same applies to $\frac{1}{3}\sqrt{3}$, and so on.

It will pay us to take a much closer look at the meaning of numbers and this we now do.

Numbers and their Definition

Most people go through life quite cheerfully without ever thinking about the precise definition of a number. In this chapter we have assumed that the idea of whole numbers is a familiar one, as of course it is.

If we reflect a little we must admit that what we do in the simplest cases is to recognise a common property possessed by several objects. For example, small children are taught to add and subtract by using beads on frames. This is a demonstration rather than a proof, and strictly speaking applies only to beads.

Can we define an integer? What is the common property possessed by all pairs of objects which we recognise at once —the property of two-ness if you like?

The study of this rather abstract subject would take us far beyond the scope of this book, but we must pursue the question of defining numbers a little further if we are to be able to prove some important propositions which will crop up later on.

Let us take for granted that we know what we mean by integers, and fractions, i.e. by rational numbers of the form p/q where p and q are integers.

We now wish to see if we can frame a definition of irrational numbers such as $\sqrt{2}$.

We know that we cannot express this as a terminating decimal, nor as a recurring one. If we could, it could be represented as a fraction.

We could of course define it by its property of having its square equal to 2, but this does not prove its existence, and we shall meet situations later when it is important to prove the existence of numbers and where a precise definition in terms of their properties would be difficult to arrive at.

We must therefore look more closely at the idea of a number such as $\sqrt{2}$.

Classes of Rational Numbers

Let us begin by dividing all rational numbers into two classes:

1. Class U—the upper class. This consists of all rational numbers whose square is greater than 2.
2. Class L—the lower class. This consists of all rational numbers whose squares are less than 2.

We shall confine ourselves to positive numbers in order to avoid confusion. It causes no loss of generality at this stage.

First note that every positive number, if rational, belongs to one or the other class.

Secondly, any member of the U class is greater than any member of the L class.

The next point is that we can find a member of U and a member of L whose difference is as small as we like.

Suppose this difference is δ. Then take a member of U and a member of L. Between two rational numbers we can insert a chain of other rational numbers whose differences are as small as we like. For example we can take their difference and divide it by 1000, or any number we wish, and so produce such a chain made up of elementary steps each equal to one thousandth of the difference.

All we need do is to form such a chain so that the steps are less than the given number δ. As we proceed along the chain of rational numbers, there must come a stage when we pass from members of L to members of U. The difference between the member of U and the member of L so defined is less than δ.

We can next show that if x is any member of L and y any member of U, we can, given δ, find an x and a y so that $2-x^2$ and y^2-2 are each less than δ.

For it is clearly possible to find an indefinitely large number of x and y so that $x+y < 4$.

Now, since δ is given we can clearly apply the previous argument to the quantity $\delta/4$. This gives us

$$y^2-x^2 = (y-x)(y+x) < 4(y-x) < \delta$$

i.e. $$(y^2-2)+(2-x^2) < \delta$$

and since each term is positive each is $<\delta$.

It now follows that L has no largest member, and U no smallest member.

For let x_m be the largest member of L, then

$$x_m^2 = 2-\varepsilon$$

But we have just proved that for any given ε we can find a member x_1 of L such that x_1^2 differs from 2 by less than ε. This means that we have found a member of L greater than the supposed largest member. This shows that L has no largest member. Similarly U has no smallest member.

Let us now take a look at the situation we have arrived at. Thinking geometrically we have two sets of points on a line which are approaching a barrier which they cannot cross. The distance between points on either side of the barrier can be made as small as we please.

We naturally ask if there could be a number which, from a geometrical point of view, corresponds to the barrier.

This number is not rational, or it would already exist as a member of U or L, i.e. it is not a member of either class. Further, since it corresponds to the barrier, it must be a number greater than any member of L and also a number less than any member of U. Let us call this number z.

Then we must have the relation

$$z^2 = 2$$

For if, for example

$$z^2 < 2 \quad \text{then} \quad z^2 = 2-\varepsilon$$

and we can by the above find a member of L greater than $2-\varepsilon$. But this means that there is a member of L greater than z, and this is impossible.

Hence, by defining classes of rational numbers as above, we have shown that there is a unique number which is not rational which forms the barrier point separating the classes if we represent rational numbers by points on a line.

In other words we have found a means of defining the irrational number $\sqrt{2}$. In the same way we can easily extend this idea to define other irrational numbers.

Suppose P and Q stand for two properties which are mutually exclusive, but one of which must be possessed by every rational number. We also suppose that numbers with the property P are less than those with the property Q. Call the numbers with property P the L class and the numbers with property Q the U class. (We have just been discussing the particular case in which the L class consisted of the rational numbers x such that $x^2 < 2$ and the U class of numbers x such that $x^2 > 2$.)

There are apparently four possibilities to consider viz.:

1. The L class has a greatest member l.
2. The U class has a least member u.
3. The L class has no greatest member, and the U class has no least member.
4. The L class has a greatest member and the U class has a least member.

It is immediately obvious that possibility 4 is ruled out. For if l and u are the greatest and least members of L and U respectively, then the number $\dfrac{l+u}{2}$ is a rational number which belongs to neither class and this contradicts our assumptions. Hence there are only three alternatives. By using exactly the same arguments as in the case of the irrational number $\sqrt{2}$ we may show that in the general case also we can find a member of L and a member of U whose difference is as small as we please.

We have thus established a process for defining a whole class of numbers which are not rational. We call these numbers irrational numbers. The totality of both rational and irrational numbers are called the class of real numbers.

Extension to Real Numbers

It is natural to ask what happens if we apply the above arguments to real numbers. Let us try and see what happens.

Take again two mutually exclusive properties P and Q, one of which must be possessed by every real number. Also let real numbers with P be less than those with Q. Form the classes L and U exactly as before except that we are now thinking of real as opposed to rational numbers.

If we now examine the four alternatives in the last section, we find that 4 is ruled out as before, but in the case of the real numbers 3 is also impossible. This is highly significant and requires proof. We now give this proof.

Take all the rational members in the L class and call this sub-class the L_1 class.

Take all the rational members of the U class and call this sub-class the U_1 class.

Consider first the L_1 class. There are only two alternatives:

(a) L_1 has a greatest member l.
(b) L_1 has no greatest member.

In case (a) then l is also the greatest member of L. For if it is not, then there is a greater member of L say l_1. But we can certainly find a rational number lying between l and l_1. This number must belong to L since it is $<l_1$ and since it is rational it belongs to L_1. But it is greater than l which is the greatest member of L_1. Thus we arrive at a contradiction and hence if l exists it is also the greatest member of L.

In case (b) L_1 has no greatest member. In this case as we have seen the classes L_1 and U_1 define an irrational number. Call this number α. Being a real number α belongs either to L or U. If it belongs to L it is the greatest member of L. For if not, there must be a greater member of L, say β.

But we can now insert a rational number between α and β. This rational number belongs to L for it is less than β. But being rational and in L it belongs to L_1. But it is greater than α. This is impossible. Hence α is the greatest member of L.

Similarly if the number α belongs to U it is the least member of U.

Hence in our classification of real numbers it may be stated that either

L has a greatest member

or U has a least member.

Both cannot occur and one *must* occur.

Thus our section of the real numbers has not generated anything new. We have reached the end.

We have dealt with the foregoing ideas in some detail, because it is essential to grasp them if what follows it to be really understood. What we have done is to expound at some length what is known as *Dedekind's Theorem*. This may be stated formally as follows:

If the real numbers can be divided into two classes L and U so that

1. Every number belongs to one or other class.
2. Each class contains at least one number.
3. Every member of L is less than every member of U.

Then there exists a number α such that all numbers less than α belong to L and all numbers greater than α belong to U. The number α may belong to either class (but not both).

The Idea of a Function

The reader will already be familiar with the idea of a function. An elementary definition is that y is a function of x if y can be found when x is given. To take a simple case, if

$$y = x^2+3 \tag{1.5}$$

then y is the function of x defined by (1.5).

Note here certain characteristics of the functional relationship (1.5).

1. The relationship is explicit, i.e. we can obtain y by calculation.
2. y is known for every value of x.
3. To each x there is a unique value of y.

This is a specially simple case of a functional relationship. There are many other examples which do not have the above properties. E.g. y is the largest prime factor of x. This only has a meaning if x is a whole number. Or y is the income in £'s per annum of x medical practitioners in the United Kingdom on 1st January 1964. y is here defined only for a finite number of integral values of x, i.e. 1, 2, 3, . . . N, where N is the number of medical practitioners in the United Kingdom cn 1st January 1964.

We mention these just to open the student's eyes to the wider possibilities of a function. However, we shall in this book be occupied with certain simple classes of functions which we shall now define.

Classes of Functions

1. *Polynomials*

Polynomials are functions of the type

$$y = a_0+a_1x+a_2x^2+\ \dots\ +a_mx^m \tag{1.6}$$

where the a's are constants.

In their simplest form they represent powers of x, such as

$$y = x^2, x^3, x^4, \text{etc.}$$

2. *Rational Functions*

A rational function is simply the ratio of two polynomials P and Q.

i.e.

$$R(x) = \frac{P(x)}{Q(x)} \tag{1.7}$$

3. *Algebraic Functions*

These are functions which can be arrived at by the general principle of formation of a rational function, but operations of root extraction are permitted.

e.g.

$$y = \sqrt{x}+\sqrt{(x+\sqrt{x})}$$

or

$$y = \left(\frac{x^2-\sqrt{x}+x}{x^5-4}\right)^{\frac{2}{3}}$$

Sometimes, when y is equal to an expression as above we can, by algebra, transform the expression into an *implicit* algebraic function involving power of y.

e.g. if

$$y = \sqrt{x}+\sqrt{(x+\sqrt{x})}$$

we can transform this into

$$y^4-(4y^2+4y+1)x = 0$$

Transcendental Functions

All functions which are not rational and not algebraic are called Transcendental Functions. This negative definition is not very satisfying and is of course extremely broad in character. We shall have to study only a few of these functions.

Examples are Trigonometric Functions such as

$$\sin x, \tan x, \cos^{-1} x, \text{etc.}$$

with which the reader must already be familiar.

Others of importance are Logarithmic Functions and the Exponential Functions where, e.g.

$$y = \log x \quad \text{or} \quad y = A\,e^x$$

where A and e are constants.

While beyond the scope of this book, mathematics abounds in other examples of transcendental functions which may arise as solutions of differential equations or in other ways. Such comprise Bessel Functions, Legendre Functions, etc., named after their discoverers.

We must now pass on to deal with some of the ideas which we said must be given precise definitions and we start with:

The Notion of Infinity

The idea of infinity will crop up frequently in more advanced work. We shall encounter, for example, infinite series and infinite integrals. We shall have to study the behaviour of functions of x when x becomes indefinitely large or, as the saying goes, 'tends to infinity'. The symbol for infinity is ∞. This symbol has no precise meaning so far, but we will give it a precise meaning when used in connection with certain phrases by defining exactly what we mean.

Let us consider first what we mean by a finite class and by an infinite (i.e. not finite) class. Take the class of whole numbers less than 1000: 1, 2, 3, . . . 999. This class is clearly finite, i.e. we can count the number in the class. The population of Great Britain is a finite class although we may not know exactly what the number of the class may be at any one time. In the same way the population of the world at any given instant is a finite, although unknown, number.

If a class is a finite one we can say with certainty that it is possible to choose a number which is larger than the number in the class. For example, in the case of the class consisting of the population of Great Britain, the number in this class is certainly less than 100,000,000, and for the class of the world population we can say with certainty that this will be less than 10,000,000,000.

If, however, we take the class of positive integers then this class is not finite, i.e. we cannot choose a number N, however large, so that it is larger than the number of positive integers. If we say, therefore, that the class of positive integers is an infinite class, we only mean that it is *not* finite.

Behaviour of Functions of Positive Integers

Let us consider the function of n when $y = f(n)$ where n is a positive integer. The class, or set, consisting of the values of $f(n)$ for various values of n is called an aggregate. We could, for example, take $y = 1/n$. Then if n takes the values 1, 2, 3 ... 100 we have an aggregate consisting of the fractions $1, \frac{1}{2}, \frac{1}{3} \ldots \frac{1}{100}$. We wish to discuss the behaviour of this function for very large values of n. It is immediately obvious that $1/n$ becomes very small when n is very large; but to say that $1/n = 0$ when $n = \infty$ is meaningless. If, however, we say that the function $y = 1/n$ tends to zero for values of n which tend to infinity, we can define precisely what we mean by this statement.

Suppose we take a number ε, however small, then we can certainly choose a number N so that function $1/n$ is less than ε for all values of $n > N$. For example, if we choose ε as ·00000001 then Y can be made less than this for values of $N \geqq 10^8$.

One way of looking at this kind of statement is for the reader to consider that he has made a bet to produce the number N with an opponent who hopes to win by making the number ε smaller and smaller. It will be obvious that no matter how small the number ε chosen by the opponent, the reader must invariably win since he can always find a number N large enough to win the wager.

If we now consider the function $y = n^2$, it is clear that this function can be made as large as we please by taking sufficiently large values of n. We may therefore say that the function $y = n^2$ tends to infinity when n tends to infinity.

For completeness one may add the function

$$y = -n^2$$

In this case we say that y tends to $-\infty$. We would usually write these statements briefly as

$$n^2 \to \infty \quad \text{or} \quad -n^2 \to -\infty \quad \text{as} \quad n \to \infty.$$

The Idea of a Limit

In considering the behaviour of a function of n we must now introduce the idea of a limit. We have seen the function $y = 1/n$ tends to zero when n tends to infinity, whereas function n^2 tends to infinity when n tends to infinity. In a more general way, some functions of n tend to definite limits when n tends to infinity and we must now find what we mean by a limit.

Definition. The function of $\phi(n)$ tends to the limit l if, given a positive number δ, however small, we can find a value n_0 so that for all values of $n \geqq n_0$

$$|\phi(n) - l| < \delta$$

Clearly n_0 depends on δ (think of the wager operation above). We may indicate this by writing $n_0(\delta)$ instead of the simple n_0.

We write this limit statement as

$$\lim_{n\to\infty} \phi(n) = l$$

We can readily extend this type of definition to functions which tends to infinity.

The function $\phi(n)$ is said to tend to $+\infty$ when n tends to infinity if, given any number Δ, however large, we can choose n_0 such that

$$\phi(n) > \Delta \quad \text{when} \quad n \geqq n_0$$

Here again n_0 depends on Δ. As an exercise, the reader should write down the definition for the function which tends to $-\infty$.

Some Properties of Limits

The *limit* of a function $\phi(n)$ is not in general equal to the *value* of the function for any value of n. It *may* be equal but this is not necessarily true; e.g. consider the function

$$\phi(n) = 1/n$$

We have seen already that

$$\lim_{n \to \infty} \phi(n) = 0$$

but there is no value of n for which $\phi(n) = 0$.

As a more complicated example, the function

$$\phi(n) = \left(\sin \frac{n\pi}{4}\right)\frac{1}{n}$$

has the limit zero when $n \to \infty$ since $\sin n\pi/4$ cannot be numerically greater than one.

There are in addition an infinite number of values of n for which the value of $\phi(n)$ is also zero. These are those values of n which are multiples of 4. For all other values of n, however, the value of $\phi(n)$ is different from its limiting value of 0.

Oscillating Functions

There exist functions of n which neither tend to infinity nor to a limit. Such functions are said to oscillate.

A simple example is

$$\phi(n) = (-1)^n$$

which takes the value $+1$ or -1 according to whether n is even or odd.

In the above example of function $\phi(n)$ is said to oscillate finitely since, although $\phi(n)$ does not tend to a limit, we can assign a number P so that

$$|\phi(n)| < P \text{ for all values of } n.$$

If there is no number P with this property, then the function $\phi(n)$ is said to oscillate infinitely.

An example of such a function is

$$\phi(n) = (-1)^n n.$$

which takes the values $-1, 2, -3, 4, -5, \ldots.$

Example. Show by writing down the first 8 terms that the function

$$\phi(n) = n\{1+(-1)^n\}$$

oscillates infinitely.

Theorems on Limits

The following four theorems enable us to manipulate functions which have limiting values.

Theorem I. If

$$\lim \phi(n) = a \quad \text{and} \quad \lim \psi(n) = b$$

then $$\lim [\phi(n)+\psi(n)] = a+b$$

The proof follows directly from the definition. It is clear that this theorem can be extended to the sum of 3 or any finite number of functions.

Theorem II. If

$$\lim \phi(n) = a \quad \text{and} \quad \lim \psi(n) = b$$

then $$\lim \phi(n)\psi(n) = ab$$

Proof. Let

$$\phi(n) = a+\phi_1(n) \quad \text{and} \quad \psi(n) = b+\psi_1(n)$$

where clearly $\lim \phi_1(n) = 0$ and $\lim \psi_1(n) = 0$

then $$\phi(n)\psi(n) = ab+a\psi_1(n)+b\phi_1(n)+\phi_1(n)\psi_1(n)$$

Whence

$$|\phi(n)\psi(n)-ab| \leqq |a\psi_1(n)|+|b\phi_1(n)|+|\phi_1(n)\psi_1(n)|$$

In the limit the R.H.S. tends to zero which proves the theorem.

Having proved the theorem for the product of two functions the extension to three or more follows immediately.

Theorem III. If $\lim \phi(n) = a$ and a is not zero

then $$\lim \frac{1}{\phi(n)} = \frac{1}{a}$$

Proof. Let $\phi(n) = a+\phi_1(n)$ where $\lim \phi_1(n) = 0$

then $$\left|\frac{1}{\phi(n)}-\frac{1}{a}\right| = \left|\frac{1}{\phi_1(n)+a}-\frac{1}{a}\right| = \frac{|\phi_1(n)|}{|a||a+\phi_1(n)|}$$

Now, since $\lim \phi_1(n) = 0$ we can clearly choose a value n_0 so that the R.H.S. is less than any quantity Δ for $n \geq n_0$.

Theorem IV. If $\lim \phi(n) = a$ and $\lim \psi(n) = b$

then $$\lim \frac{\phi(n)}{\psi(n)} = \frac{a}{b}$$ provided b is not zero.

The proof follows immediately for theorems II and III.

Functions which Increase Steadily with n

A function $\phi(n)$ is said to increase steadily with n if

$$\phi(n+1) \geqq \phi(n) \quad \text{for all values of } n. \tag{1.8}$$

Such functions have some extremely important properties.

Theorem. If $\phi(n)$ increases steadily with n then either

$$\phi(n) \to \text{limit} \quad \text{or} \quad \phi(n) \to \infty$$

A rigorous proof of this theorem is too difficult to give here. The important point to grasp is that if we *cannot* find a number K with the property that

$$\phi(n) < K \quad \text{for all values of } n > n_0$$

then we can assert that

$$\lim \phi(n) \to \infty$$

If, on the other hand, there *is* such a number K such that

$$\phi(n) < K \quad \text{for all } n > n_0$$

then we can assert that

$$\lim_{n \to \infty} \phi(n) \qquad \text{exists, and further that}$$

$$\lim \phi(n) \leqslant K$$

Note that this theorem applies to *steadily increasing functions.*

The importance of this theorem is that it enables us to prove the existence of a limit without knowing what the value of the limit may be.

It is easy to see by applying similar reasoning that a function $\phi(n)$ which *decreases* steadily must tend to a limit a or $-\infty$.

The Limit of x^n as $n \to \infty$

This is a very important case.

If $\quad x = 1 \quad x^n = 1$ always and $\lim \phi(n) = 1$
If $\quad x = 0 \quad x^n = 0$ always and $\lim \phi(n) = 0$

Now let x be positive, then

$$\phi(n+1) = x\phi(n)$$

and hence if $x > 1$, $\phi(n)$ is a steadily increasing function and if $0 < x < 1$, $\phi(n)$ is a steadily decreasing function. In the case of $x > 1$, $\phi(n) \to \infty$ because if $\phi(n) \to l$ then

$$\lim \phi(n+1) = \lim x\phi(n) = xl$$

but this implies that $l = xl$ and, since both x and l are greater than one, this is impossible.

Hence for $\qquad x > 1 \qquad \phi(n) \to \infty$

If, however, $0 < x < 1$ then x^n is a decreasing function of n and so must lead to a limit l or to $-\infty$. Since x is assumed positive the last alternative is excluded and we know that $\lim \phi(n)$ exists. Let the limit be l. By applying the reasoning above

$$l = xl$$

But since x is not unity we must have

$$l = 0$$

Hence if $\qquad 0 < x < 1, \qquad \lim \phi(n) = 0.$

We must now consider the cases in which x is negative. If $-1 < x < 0$ then if we put $x = -y$ we have $0 < y < 1$ and from the previous discussion, $\lim y^n = 0$ and hence

$$\lim x^n = 0.$$

If $x = -1$, it is obvious that x^n oscillates finitely and if $x < -1$, then clearly x^n oscillates infinitely.

The Upper Bound of an Aggregate

We shall now use Dedekind's Theorem to prove some very important results.

Consider first a function $\phi(n)$ where n is an integer.

If we can show the existence of a number K, such that

$$\phi(n) < K \tag{1.9}$$

for all values of n, then we say that $\phi(n)$ is bounded above. Clearly, if one such number exists, then there are infinitely many others (any number $> K$).

We wish to show that there is a *least* which we denote by M and we call M the Upper Bound of the aggregate $\phi(n)$.

To prove that M exists, take two classes of real numbers: L class consisting of numbers which are exceeded by some members of the aggregate $\phi(n)$, U class consisting of numbers which are not exceeded by *any* member of the aggregate $\phi(n)$.

Clearly the conditions of Dedekind's Theorem are satisfied. Both classes exist and every member of the L class is less than every member of the U class.

Hence the classification defines a real number M which is the one we are looking for.

Note that by definition M is not less than *any* member of the aggregate $\phi(n)$ but that any number $< M$, say $M-\varepsilon$, is exceeded by *at least one* member of $\phi(n)$.

In precisely the same way we can prove the existence of a lower bound m for an aggregate which is bounded below. In certain cases an aggregate may be bounded both above and below and in this case

$$m < \phi(n) < M \qquad \text{for } \textit{all} \text{ values of } n.$$

Behaviour of Functions of a Continuous Variable

So far we have been considering the behaviour of functions involving an integer n which tended to infinity or to some other value. Clearly we must extend our ideas to cover the behaviour of functions of x where now x means a continuous real variable. Just as the integer n could be thought of as tending to infinity so the continuous variable x can be thought of as tending to infinity. We start with a definition.

I. The function $f(x)$ is said to tend to the limit l as x tends to ∞ if, when a positive number δ, however small, has been chosen, it is possible to choose a number $x_0(\delta)$ such that for all values of $x \geqq x_0$

$$|f(x)-l| < \delta$$

When this is so we write

$$\lim_{x\to\infty} f(x) = l$$

II. If when a number Δ, however large, is given we can choose $x_0(\Delta)$ such that

$$f(x) > \Delta$$

for all values of $x \geqq x_0$ then we can say that $f(x)$ tends to ∞ with x and write

$$f(x) \to \infty$$

The definition for the case when $f(x) \to -\infty$ is left to the reader.

Finally, the function $f(x)$ may oscillate finitely or infinitely when neither of the definitions I or II is satisfied.

Limits of $f(x)$ when $x \to 0$

Before giving a formal definition suppose we have a function

$$y = \phi(x)$$

and that we know the behaviour of y when $x \to \infty$, e.g.

$$\lim_{x\to\infty} \phi(x) = l$$

Now imagine x replaced by the substitution

$$z = \frac{1}{x}$$

Then when $x \to \infty$ $z \to 0$ and we may now think of y merely as a function of z,

$$y = \psi(z)$$

and study the behaviour of y as $z \to 0$.

There is one catch here. $z \to 0$ by *positive* values and we signify this by writing

$$\lim_{z \to +0} \psi(z) = l$$

If we were considering the behaviour of $y = \phi(x)$ when $x \to -\infty$, and if the limit exists and equals l', then by making the substitution for z above, we would now have

$$\lim_{z \to -0} \psi(z) = l'$$

We do not wish to weary the student with too many definitions but the following one is important.

Definition. If when any positive number δ is given, however small, we can choose $y_0(\delta)$ so that for all values of y different from zero but numerically less than $y_0(\delta)$, we have

$$|f(y) - l| < \delta$$

then we may say $f(y)$ tends to the limit l as y tends to zero and write this

$$\lim_{y \to 0} f(y) = l$$

Note that y may tend to zero by *either positive or negative values.*

We must stress an important point immediately.

The statement

$$\lim_{x \to 0} f(x) = l$$

is *not* a statement about the value of $f(x)$ when $x = 0$; it is a statement about its value when x is *very nearly* equal to 0.

For example, let $$y = \frac{x}{x}$$

This function is equal to 1 for all values of x save $x = 0$. It is meaningless when $x = 0$.

Hence here $$\lim_{x\to 0} y = 1$$

We do not know what the value of y may be for $x = 0$. There is, of course, no reason why the value for $x = 0$ and the limit for $x \to 0$ should not sometimes be equal. This is frequently so but it is not *necessarily* so.

Limits of a Function of x when x Tends to a Value α

Let $$\lim_{x\to 0} f(x) = l$$

Then if we put $$x = y - \alpha$$

$$f(x) = f(y-\alpha) = \phi(y)$$

Clearly when $x \to 0$, $y \to \alpha$ and so

$$\lim_{y\to\alpha} \phi(y) = l$$

We can formalise this statement as follows:
If, given δ, however small, we can find $\varepsilon(\delta)$ such that

$$|\phi(y) - l| < \delta$$

provided that $$0 < |y-\alpha| \leqq \varepsilon(\delta)$$

then $$\lim_{y\to\alpha} \phi(y) = l$$

The Upper and Lower Bounds of a Function of a Continuous Variable

Suppose $f(x)$ has the property that there is a number K such that $f(x) < K$ for all x.

Then we say that $f(x)$ is bounded above.

We may now show exactly as on page 20 that there exists a number M which is the Upper Bound of $f(x)$.

$f(x)$ never exceeds M but any number $< M$ is exceeded by at least one value of $f(x)$.

Similarly we can prove that if $f(x)$ is bounded below, that there is a number m which is the Lower Bound of $f(x)$.

Continuous Functions

The theorems we have just proved for limits enable us to define what we mean by a continuous function.

Definition. A function $f(x)$ defined for the interval $a \leqq x \leqq b$ is said to be continuous at a point c in the interval if

$$\lim_{x \to c} f(x) = f(c)$$

Note that we assume that $f(c)$ has a definite value and that x may tend to the value c from either side. Note also that in the case of a continuous function the *value* of $f(x)$ for $x = c$ and the *limit* of $f(x)$ when $x \to c$ are equal.

A function $f(x)$ is said to be continuous in an interval if it is continuous for all values of x in the interval.

From a graphical point of view, if we were to plot the graph of $y = f(x)$ it would appear as an unbroken line.

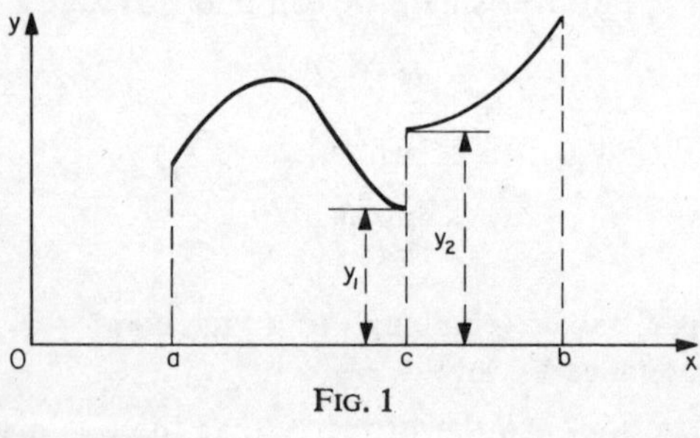

FIG. 1

There can exist cases such as in Figure 1, where the function $f(x)$ is discontinuous at the point $x = c$. In our previous terminology

$$\lim_{x \to c-0} f(x) = y_1 \quad \text{and} \quad \lim_{x - c+0} f(x) = y_2(y_1 \neq y_2)$$

We shall now prove the fundamental property of a continuous function, viz.

If $f(x)$ is a function which is continuous over a range from a to b and if

$$f(a) = y_0 \quad \text{and} \quad f(b) = y_1$$

then if y is a number between y_0 and y_1, there is some value of x such that

$$f(x) = y$$

i.e. as x varies from a to b, $f(x)$ must assume at least once every value between y_0 and y_1.

The converse is not true. The function whose graph is shown in Figure 1 assumes every value of y in the interval, but it is not continuous.

Let $y = f(x)$ be a continuous function and $f(a) = y_0$, $f(b) = y_1$. Let y be a value such that

$$y_0 < y < y_1$$

and we assume that this relation holds for all values of y in the interval (a, b).

Then there is some value ξ_1 of x such that

$$f(\xi_1) = y$$

This is illustrated in Figure 2.

Since $f(x)$ is continuous there are certainly some values of x to the right of $x = a$ for which $f(x) < y$.

Similarly there are some values of x to the left of $x = b$ for which

$$f(x) > y.$$

Divide all numbers ξ in the range (a, b) into two classes:

L Class all numbers ξ with the property

$$f(\xi) < y$$

U Class all numbers ξ with the propert

$$f(\xi) \geqq y$$

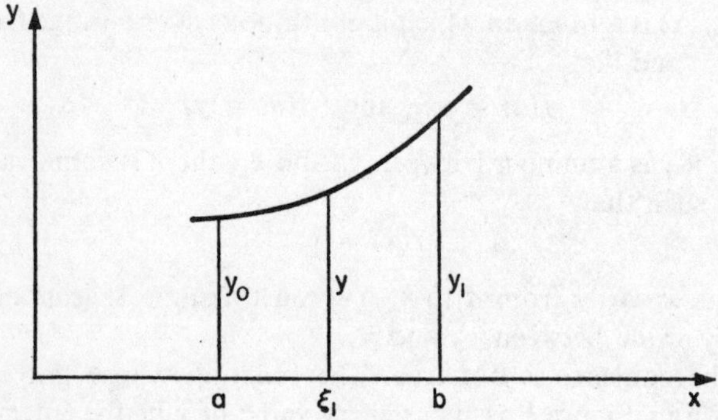

FIG. 2

Then the conditions of Dedekind's Theorem are satisfied and the resulting section defines a number. Call this number ξ_1.

Then we assert that $\quad f(\xi_1) = y$

This assertion requires to be proved. This we now do.

Suppose $\quad f(\xi_1) > y \quad$ say $\quad f(\xi_1) = y+\varepsilon$

In this case ξ_1 is in U.

We can certainly find a value ξ' such that $f(\xi') < y$ and hence, deduce that

$$f(\xi_1)-f(\xi') > \varepsilon$$

Now no matter how near ξ' may be to ξ_1 the above conditions hold, and this violates the condition that $f(\xi)$ is continuous at ξ_1. Hence $f(\xi_1) \not> y$ and in the same way we can show that $f(\xi_1) \not< y$.

Thus $f(\xi_1) = y$ as we had to prove.

This means that as x varies from a to b, y must assume at least once every value between $f(a)$ and $f(b)$.

For example, a polynomial is continuous except where x tends to infinity.

A rational function of x is continuous except for values of x which make the denominator zero.

An important case is the following.

If $y = f(x)$ is continuous for $x = x_0$, and if $z = \phi(y)$ is continuous at y for $y = y_0$ then the function

$$z = \phi[f(x)]$$

will be continuous at $x = x_0$.

For we have given that

$$\lim_{x \to x_0} f(x) = f(x_0) \quad \text{and} \quad \lim_{y \to y_0} \phi(y) = \phi(y_0)$$

It is worth while to give a formal proof.

We have $\quad |z-z_0| = |\phi[f(x)] - \phi[f(x_0)]|$

Now since $\phi(y)$ or $\phi[f(x)]$ is a continuous function, we can choose a number δ such that given $\varepsilon > 0$,

$$|\phi[f(x)] - \phi[f(x_0)]| < \varepsilon \tag{1.10}$$

provided $\quad |f(x)-f(x_0)| < \delta(\varepsilon) \qquad (1.11)$

But since $f(x)$ is continuous we can, given any number $\delta(\varepsilon)$ choose a value ρ such that

$$|f(x)-f(x_0)| < \delta(\varepsilon) \tag{1.12}$$

provided that $\quad |x-x_0| < \rho[\delta(\varepsilon)] \qquad (1.13)$

Hence if $|x-x_0|$ satisfies (1.13) then (1.11) is true and hence (1.10) is true, which proves the theorem.

The Upper and Lower Bounds of a Continuous Function

Let $f(x)$ be a function continuous in an interval (a, b). Then we have two important theorems.

Theorem I. $f(x)$ is bounded in the interval (a, b).

We can certainly find an interval (a, ξ) in which $f(x)$ is bounded. For by definition of continuity we can make $f(a+\delta)-f(a) < \varepsilon$ and ε is given and this entails that $f(x)$ is boundea in tne interval $(a, a+\delta)$.

So we now use Dedekind's Theorem to clarify the points ξ. ξ is a member of L if $f(x)$ is bounded in the interval (a, ξ). ξ is a member of U if $f(x)$ is not.

We have seen that L exists and we shall prove that U does not.

Suppose U does exist. Then we have defined a number α which is the barrier between the classes.

But $f(x)$ is continuous in (a, b). Hence given ε we can choose δ so that

$$f(x+\delta)-f(x) < \varepsilon$$

or

$$f(x-\delta)-f(x) < \varepsilon$$

and if we choose β as a particular value of x then $f(x)$ is bounded in the region $(a, \beta+\delta)$.
This is a contradiction. Hence the U class does not exist and so $f(x)$ is bounded throughout (a, b).

Theorem II. If $f(x)$ is continuous in (a, b) then it has an upper bound M and a lower bound m in the interval and it assumes both values *at least once* in the interval.

Proof. Let ξ be any number in (a, b).

Then if M is the upper bound of $f(x)$ in (a, b) and $M(a, \xi)$ the upper bound in the interval (a, ξ) then $M(a, \xi)$ is a steadily increasing function of ξ and never exceeds M.

$\therefore$ Form a section

$$\text{Class L} \quad M(a\ \xi) < M$$
$$\text{Class U} \quad M(a\ \xi) = M$$

Let β be the number so defined.

$$M(a, \beta-\varepsilon) < M$$
$$M(a\,\beta+\varepsilon) = M$$

where $$M(\beta-\varepsilon, \beta+\varepsilon) = M$$

But this means that in the interval $(\beta-\varepsilon, \beta+\varepsilon)$ there is a value of x so that $f(x) > M-\delta$ where δ is given, and since (x) is continuous this entails $f(\beta) = M$.

Examples

Let us examine the following limits:

1. $\lim\limits_{n\to\infty} \dfrac{10^7}{\sqrt{n}}$. Here the limit is clearly zero.

2. $\lim\limits_{n\to\infty} \dfrac{\sqrt[3]{n}}{10^6}$. Here the limit is clearly infinity.

3. $\lim\limits_{n\to\infty} \dfrac{\sin n\theta}{n}$. $\left|\dfrac{\sin n\theta}{n}\right| < \dfrac{1}{n}$ and so the limit is zero.

4. $\lim\limits_{n\to\infty} \dfrac{a \sin^2 n\theta + b \cos^2 n\theta}{\sqrt{n}}$, the function is $< \dfrac{a+b}{\sqrt{n}}$ and so tends to zero.

5. $\lim\limits_{n\to\infty} \left\{\dfrac{1}{n^2}+\dfrac{2}{n^2}+\dfrac{3}{n^2}+ \ldots + \dfrac{n}{n^2}\right\}$, the expression in brackets $= \dfrac{\frac{1}{2}n(n+1)}{n^2}$ which tends to $\dfrac{1}{2}+\dfrac{1}{2n} \to \dfrac{1}{2}$.

6. $\lim\limits_{n\to\infty} n.(-1)^n = -1, 2, -3, 4 \ldots,$ this oscillates infinitely.

7. $\lim\limits_{n\to\infty} n^2\{1+(-1)^n\} = 0, 8, 0, 32 \ldots,$ this oscillates infinitely.

8. $\lim\limits_{n\to\infty}$ smallest prime factor in N.

This is 2 when N is even but N when N is prime. Hence this function oscillates infinitely.

9. Let us examine the behaviour of the function $\lim_{n\to\infty} nx^n$ for various values of x.

For $x = 0$ clearly $\lim nx^n = 0$
For $x = 1$ clearly $\lim nx^n \to \infty$
For $x > 1$ clearly $\lim nx^n \to \infty$

For $x < 1$, let us write $\phi(n) = nx^n$

then
$$\frac{\phi(n+1)}{\phi(n)} = \frac{n+1}{n}x = \left(1+\frac{1}{n}\right)x$$

Hence we can find a number n_0 such that for $n > n_0$

$$\left(1+\frac{1}{n}\right).x < K < 1$$

But if
$$\frac{\phi(n+1)}{\phi(n)} < K$$

$$\frac{\phi(n+1)}{\phi(1)} < K^n$$

and since
$$K < 1 \quad \lim_{n\to\infty} K^n = 0 \quad \text{and so} \quad \lim_{n\to\infty} nx^n = 0$$

if $x < 1$.

10. We must now consider a special case of great importance. It is

$$\lim_{n\to\infty}\left(1+\frac{1}{n}\right)^n$$

We shall show that this limit exists and that it lies between 2 and 3.

$$\left(1+\frac{1}{n}\right)^n = 1+\frac{n}{n}+\frac{n(n-1)}{1.2}\frac{1}{n^2}+ \ldots +\frac{n(n-1)\ldots 1}{1.2\ldots n}\frac{1}{n^n}$$

$$= 1+1+\frac{1}{1.2}\left(1-\frac{1}{n}\right)+\frac{1}{1.2.3}\left(1-\frac{1}{n}\right)\left(1-\frac{2}{n}\right)+\ \dots \qquad (1.14)$$

In this sum all the terms are positive and so $(1+1/n)^n$ is an increasing function of n and so must tend to infinity or to a limit.

But,

$$\left(1+\frac{1}{n}\right)^n < 1+1+\frac{1}{1.2}+\frac{1}{1.2.3}+\ \dots\ +\frac{1}{1.2.3\ \dots\ n}$$

and so $< 1+1+\frac{1}{2}+\frac{1}{2^2}+\ \dots\ +\frac{1}{2^{n-1}}$

or $< 1+2$

i.e. < 3

Hence (1.14) cannot tend to infinity and has a limit < 3. The limit is obviously > 2. This proves our result. In fact the value of the limit is a transcendental number like π

Its value is 2·71828 . . .

It is an extremely important number in higher analysis and is always denoted by the letter e.

11. As a final exercise on limits let us find $\lim n^{1/n}$ when $n \to \infty$

We have just proved that

$$\left(1+\frac{1}{n}\right)^n < n \quad \text{if} \quad n \geqq 3$$

Taking this for granted, we have

$$(n+1)^n < n^{n+1}, \text{ so } (n+1)^{1/n+1} < n^{1/n}$$

Hence the function $n^{1/n}$ decreases steadily. But it is always > 1 so it must tend to a limit > 1.

Suppose this limit is l where $l > 1$, so $n^{1/n} \to l > 1$. Then $n > l^n$ for sufficiently large values of n. This is certainly untrue, for

$$\lim_{n\to\infty} \frac{n}{l^n} = 0 \quad \text{if} \quad l > 1$$

Hence
$$\lim_{n\to\infty} n^{1/n} = 1$$

CHAPTER TWO

DIFFERENTIAL CALCULUS—REVISION

Definitions

We shall assume that the reader is already familiar with the ideas of the differential calculus from an elementary point of view. He will doubtless have acquired some manipulative skill and be aware of the standard formulae employed in finding differential coefficients.

A stumbling block to a fuller understanding of the meaning of a derivative when the idea is first encountered is, of course, the fact that it involves the idea of a limit.

The well known expression for the derivative of $f(x)$, viz.,

$$\frac{d}{dx}[f(x)] = \lim_{l \to 0} \frac{f(x+l)-f(x)}{l} \tag{2.1}$$

speaks for itself.

The reader who has carefully studied Chapter One will have no difficulty with such ideas and we shall make full use of them in a revision of the basic ideas.

Consequences of Equation 2.1

If the limit on the R.H.S. of equation (2.1) exists then $f(x)$ is a continuous function.

This is obvious because the existence of the limit in (2.1) implies that

$$\lim_{l \to 0} f(x+l) = f(x)$$

For, if not, then

$$\{\lim_{l \to 0} f(x+l)\} - f(x) = \alpha$$

when α is not zero, and then

$$\lim_{l\to 0}\frac{f(x+l)-f(x)}{l} = \lim_{l\to 0}\frac{\alpha}{l}$$

which does not exist.

The converse is not true. i.e. a function may be continuous but need not necessarily have a derivative at every point. This is easily seen by a function which consists of two straight lines as in Figure 3. The derivative at P has *two*

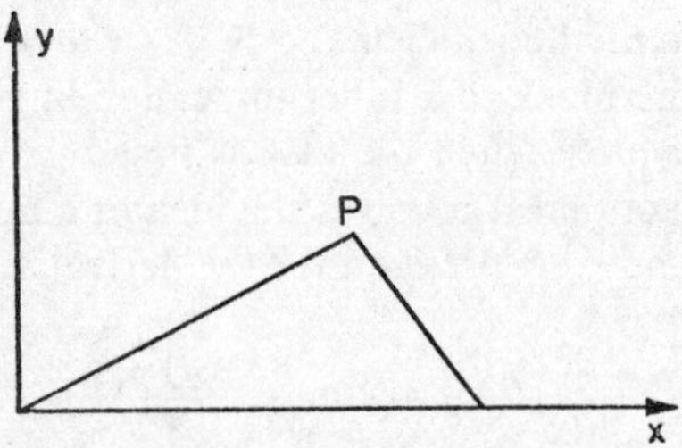

FIG. 3

values depending on how we approach the limit. In the definition (2.1) we of course assume that l may approach zero by either positive or negative values.

We remind the reader that there is a geometrical illustration of a derivative.

The derivative of a function $y = f(x)$ can be given the geometrical meaning that it is the *slope* of the curve $y = f(x)$. In other words, if we draw a tangent to the curve at some point $x = x_0$ then $\tan\psi = f'(x_0)$ where ψ is the angle made by the tangent and the x axis.

A Word of Caution

We must point out that when the tangent is parallel to the axis of y, the derivative is discontinuous. In this case $\psi = 90°$ and $\tan\psi \to \infty$ as $\psi \to 90°$, and also the function

$f(x)$ may be such that $f(x+l)$ may be undefined for positive values of l. This is illustrated by Figure 4.

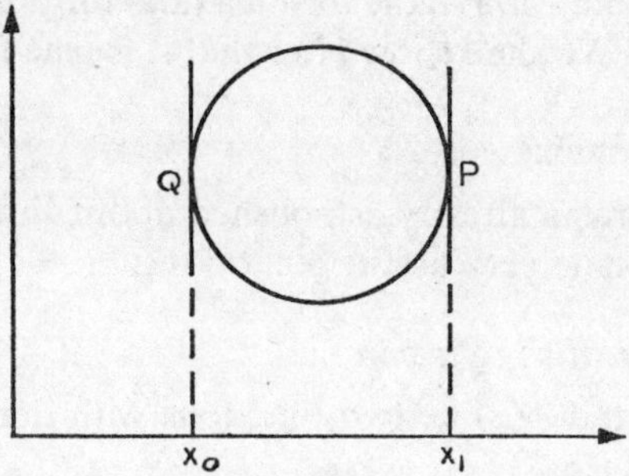

FIG. 4

If we consider the case of a circle at the point P, for example, the function is such that it is undefined for $x > x_1$ and two valued for $x < x_1$. The slope of the tangent at P therefore is such that the value of $\tan \psi$ tends to $+\infty$ or to $-\infty$ depending on whether we approach the limit from below the point P or above.

A. Standard Formula

Starting from the definition (2.1) the reader should have no difficulty in proving the well known formula for the derivative of x^n, viz.

$$\frac{dx^n}{dx} = nx^{n-1} \tag{2.2}$$

For, by the binomial theorem

$$(x+h)^n = x^n + nx^{n-1}h + \frac{n(n-1)}{1.2}x^{n-2}h^2 + \ldots + h^n$$

and hence

$$\frac{(x+h)^n - x^n}{h}$$

$$= nx^{n-1} + \text{terms involving } h \text{ or higher powers of } h.$$

These terms vanish when we take the limit for $h \to 0$. Hence the formula (2.2) is established.

We have only *proved* the formula (2.2) for positive integral values of n. We shall show later that it is true for *all* values.

General Formulae

The theorems already established about limits enable us to deduce some very useful general results.

(1) *The derivative of a sum*

Let $f(x)$ and $\phi(x)$ be two functions with derivatives $f'(x)$ and $\phi'(x)$, then

$$\frac{d}{dx}[f(x)+\phi(x)] = f'(x)+\phi'(x) \tag{2.3}$$

This can be deduced directly from the definition and clearly extended to the sum of any (finite) number of differential functions.

(2) *The derivative of a product*

Let $f(x)$ and $\phi(x)$ be two differentiable functions

$$F(x) = f(x).\phi(x)$$

Then
$$F'(x) = f(x)\phi'(x)+f'(x)\phi(x)$$

The proof is straightforward, for,

$$F'(x) = \lim_{h\to 0}\frac{f(x+h)\phi(x+h)-f(x)\phi(x)}{h}$$

$$= \lim_{h\to 0}\left\{f(x+h)\frac{[\phi(x+h)-\phi(x)]}{h} + \phi(x)\frac{[f(x+h)-f(x)]}{h}\right\}$$

$$= f(x)\phi'(x)+\phi(x)f'(x) \tag{2.4}$$

We may deduce immediately that

$$\frac{d}{dx}(C.f(x)) = Cf'(x) \quad \text{if } C \text{ is a constant}$$

(3) *Derivative of a quotient*

As formerly, it is simpler to derive the expression for a reciprocal and then use (2) above.

Let $$\phi(x) = \frac{1}{f(x)}$$

then $$\phi'(x) = -\frac{f'(x)}{[f(x)]^2} \tag{2.5}$$

provided, of course, that $f(x)$ is not zero for the value of x under consideration.

Proof.

$$\phi'(x) = \lim_{h\to 0}\left\{\frac{1}{h}\left(\frac{1}{f(x+h)} - \frac{1}{f(x)}\right)\right\}$$

$$= \lim_{h\to 0}\left\{\frac{f(x)-f(x+h)}{h} \times \frac{1}{f(x)f(x+h)}\right\}$$

$$= \frac{f'(x)}{[f(x)]^2}$$

Passing now to the case

$$\phi(x) = \frac{f(x)}{F(x)}$$

we have at once

$$\phi'(x) = \frac{f'(x)F(x) - F'(x)f(x)}{[F(x)]^2} \tag{2.6}$$

Finally, we have the derivative of a function of a function.

Let $\phi(x) = F(y)$ where $y = f(x)$

or $\phi(x) = F[f(x)]$

Then we shall prove the well known and very important result.

$$\phi'(x) = F'(f(x)) \times f'(x) \tag{2.7}$$

Let h be a small but finite quantity, then, if $y = f(x)$, we may write

$$f(x+h) = f(x)+r$$

because $f(x)$ is continuous and, of course, $\lim_{h \to 0} r = 0$.

Note also that

$$\lim_{h \to 0} \frac{r}{h} = f'(x)$$

$$\phi'(x) = \lim_{h \to 0} \frac{\phi(x+h)-\phi(x)}{h} = \lim_{h \to 0} \frac{F[f(x+h)]-F[f(x)]}{h}$$

$$= \lim_{h \to 0} \frac{F(y+r)-F(y)}{r} \cdot \frac{r}{h}$$

$$= F'(y) \times f'(x) = F'f(x) \times f'(x)$$

since $r \to 0$ when $h \to 0$.

This last theorem is often expressed by the formula

$$\frac{dy}{dx} = \frac{dy}{du}\frac{du}{dx} \tag{2.8}$$

where $y = y(u)$ and $u = u(x)$.

This is easier to remember but it has a tendency to look obvious. We cannot 'cancel' the *du*'s since, taken by themselves, they have no meaning.

Inverse Functions

Suppose $y = f(x)$ but we are given instead the inverse

$$x = \phi(y)$$

then it is true that

$$\frac{dy}{dx} = \frac{1}{dx/dy} \tag{2.9}$$

but here again the 'obvious' nature of this formula should not mask the fact that it requires proof.

If $$y = f(x) \quad \text{then} \quad f(x+h) = y+k$$

where $k \to 0$ when $h \to 0$ and also if the inverse function is

$$x = \phi(y), \quad \text{then} \quad x+h = \phi(y+k)$$

but, $$f'(x) = \lim_{h\to 0} \frac{f(x+h)-f(x)}{x+h-x}$$

$$= \lim_{k\to 0} \frac{y+k-y}{\phi(y+h)-\phi y} = 1/\phi'(y)$$

which proves the theorem.

Derivative of Common Functions

We shall use our previous classification.

(1) *Polynomials*

If $$f(x) = a_n x^n + a_{n-1}x^{n-1} + \ldots + a_0$$
$$f'(x) = na_n x^{n-1} + (n-1)a_{n-1}x^{n-2} + \ldots + a_1$$

(2) *Rational Functions*

Let $$F(x) = \frac{P(x)}{Q(x)}$$

then $$F'(x) = \frac{P'(x)Q(x) - Q'(x)P(x)}{[Q(x)]^2}$$

This rule can lead to some complicated algebra so it is better tactics to do the algebra first by reducing the function $F(x)$ by finding the factors of the function Q. It is then possible to express the function $R(x)$ in the form

$$R(x) = S(x) + \text{a series of terms of the form}$$

$$\frac{a_1}{x-\alpha_1} + \frac{b_1}{(x-\alpha_1)^2} + \ldots$$

and $$\frac{a_2}{x-\alpha_2}+\frac{b_2}{(x-\alpha_2)^2}+\ldots, \text{etc.,}$$

where $S(x)$ is a polynomial and $\alpha_1, \alpha_2, \ldots$ are the factors of $Q(x)$.

The derivative of $R(x)$ may now be written down

$$R'(x) = S'(x)-\frac{a_1}{(x-\alpha_1)^2}-\frac{2b_1}{(x-\alpha_1)^3}-\ldots$$
$$-\frac{a_2}{(x-\alpha_2)^2}-\frac{2b_2}{(x-\alpha_2)^3}-\ldots, \text{etc.}$$

Algebraic Functions

All we require to do is to apply the theorem for the derivative of a function of a function (p. 38) and proceed from there.

As an example. Let $y = x^{p/q}$ and $x = z^q$ where $y = z^p$ and p and q are integers,

then $$\frac{dy}{dx} = \frac{dy}{dz}\div\frac{dx}{dz} = pz^{p-1}\div qz^{q-1}$$

$$= \frac{p}{q}z^{p-q} = \frac{p}{q}x^{p/q-1}$$

This provides a proof of the fact that the formula

$$\frac{dx^n}{dx} = nx^{n-1}$$

is true for rational values of n.

The student should now show that it is true for negative integral values of n.

Implicit Algebraic Functions

Suppose we have an implicit relation

$$ax^2+by^2+cy^3+\lambda xy = 0$$

we merely differentiate each term, getting

$$2ax+2by\frac{dy}{dx}+3cy^2\frac{dy}{dx}+\lambda y+\lambda x\frac{dy}{dx} = 0$$

and solving we obtain

$$\frac{dy}{dx} = \frac{-(2ax+\lambda y)}{\lambda x+2by+3cy^2}$$

Examples

Differentiate the following:

1. $5x^4+7x^3+2x-3$
2. $(x-4)(x+5)(x-1)$
3. $\frac{x}{1-x^2}$
4. $\sqrt{x}-\frac{1}{\sqrt{x}}$
5. $\sqrt{(\sqrt{x}+x)}$
6. $\sqrt{\frac{1-x}{1+x}}$
7. $\frac{(3x+2)^3}{(2x-1)^2}$
8. Find $\frac{dy}{dx}$ where $ax^2+2hxy+by^2 = 0$

The functions which still remain to be studied are

Transcendental Functions

We shall start with the trigonometric functions with which the student must already be familiar.

The usual definition of these functions in an elementary treatment is based on geometrical ideas. Referring to Figure 5, we define

$$\sin\theta = \frac{PR}{OP}$$

$$\cos\theta = \frac{OR}{OP} \tag{2·10}$$

$$\tan\theta = \frac{PR}{OR}$$

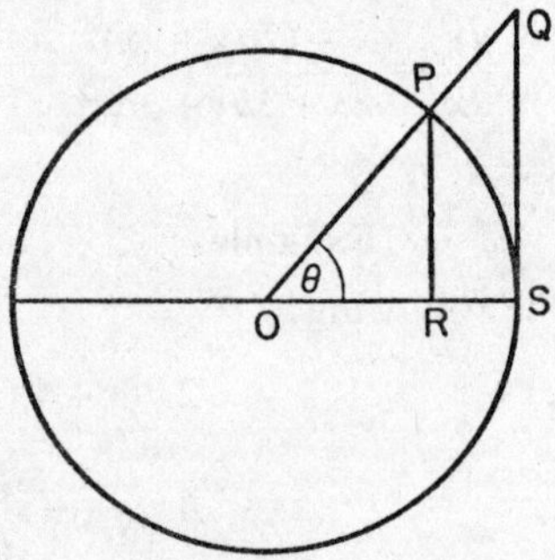

FIG. 5

Note that with the usual sign convention we may deduce from the above that

$$\begin{aligned}\sin(-\theta) &= -\sin\theta \\ \cos(-\theta) &= \cos\theta \\ \tan(-\theta) &= -\tan\theta\end{aligned}$$

By purely geometrical arguments, to be found in any elementary text book on trigonometry, we may prove the sum and difference formulae, viz.

$$\sin(x+y) = \sin x\cos y + \cos x\sin y \tag{2.11}$$

$$\cos(x+y) = \cos x\cos y - \sin x\sin y \tag{2.12}$$

and the associated special cases

$$\begin{aligned}\sin 2x &= 2\sin x\cos x \\ \cos 2x &= \cos^2 x - \sin^2 x = 1 - 2\sin^2 x\end{aligned}$$

We may also, if rather less rigorously, deduce the expression for the area of the sector of a circle of radius r which subtends an angle of amount (in radians) at the centre, viz.

$$\text{area} = \tfrac{1}{2}r^2\theta \tag{2.13}$$

Although these expressions do define certain functions in a geometrical way, they leave rather vague the concept of what we mean precisely by the variable θ.

What exactly do we mean by an angle such as θ?

We have a kind of intuitive idea that it is associated with the area bounded by the lines OP and OS and the arc of the circle SP. Is there, however, some way of precisely measuring this quantity? [A moment's reflection will suffice to convince the reader that the reply 'With a protractor', merely begs the question.]

One way is to take the unit of length so that the radius of the circle is unity. Then we could theoretically measure the angle by measuring the length of the arc SP.

If we were to do this, it would follow at once that

$$\theta = 1 \quad \text{when} \quad SP = OS = 1$$

and that $\theta = 2\pi$

when the vector OP has swept right round the circle.

Note here that with this system, the number of units of angle in a complete circle is not a rational multiple of the length of the radius.

This definition of an angle may look all right, but it does assume that we know how to measure the lengths of curved lines in general or at least of arcs of circles in particular. Later on, when we discuss integration, we shall show how this can be done. For the present we only invite the reader to consider the difficulty.

Another way of measuring the angle would be to take again a circle of unit radius and to define the measure of the angle θ as

$$\text{measure of } \theta = \tfrac{1}{2}(\text{area of sector } OPR) \tag{2.14}$$

Here again we assume that we are able to measure areas bounded by curves and rules for doing this will appear when we study integration.

The reader may reasonably ask what is wrong with the normal practical method of measuring an angle by dividing the angle subtended by a full circle into 360 parts and calling each part a degree.

We have here the difficulty that if the radius of the circle is a rational number, then we cannot divide the circumference into 360 equal parts. Naturally from a purely practical point of view we could do so to a degree of accuracy amply sufficient for such purposes as surveying, etc.

It may be worth mentioning that the number 360 is itself based on a false assumption. The early astronomers, with their imperfect instruments, thought that the earth took 360 days to circle the sun, and hence divided a circle into 360 equal parts. One wonders what they would have done had they known that the number was nearly $365\frac{1}{4}$.

We shall return to this subject later on. For the present we hope that the reader has realised that such an apparently simple concept as an angle is not nearly so simple as it looks.

Needless to say there are other ways of defining the trigonometric functions which we shall discuss later. Some of them evade the difficulties mentioned above.

However, in order to push on and develop the theory of differentiation of trancendental functions we propose to assume the validity of the geometrical definitions for the time being.

We must proceed to prove one very important result:

$$\lim_{x\to 0}\left[\frac{\sin x}{x}\right] = 1 \qquad (2.15)$$

Although the proof must be taken with reservations at this stage because it involves the concept of measuring an area, it will be shown later to be capable of rigorous treatment. Referring to Figure 6,

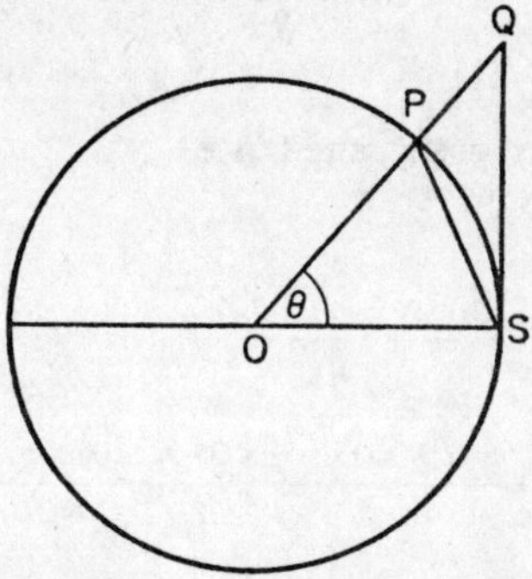

FIG. 6

area of ΔOPS < area of sector OPS < area of ΔOQS. Using our definitions and taking a unit circle, we can express this

$$\tfrac{1}{2}\sin\theta < \tfrac{1}{2}\theta < \tfrac{1}{2}\tan\theta$$

i.e. $$1 < \frac{\theta}{\sin\theta} < \frac{1}{\cos\theta} \quad \text{or} \quad 1 > \frac{\sin\theta}{\theta} > \cos\theta$$

and hence $$0 < 1 - \frac{\sin\theta}{\theta} < 2\sin^2\frac{\theta}{2} < 2(\tfrac{1}{2}\theta)^2$$

i.e. $$0 < 1 - \frac{\sin\theta}{\theta} < \frac{\theta^2}{2}$$

so, clearly, $$\lim_{\theta\to+0}\left(1 - \frac{\sin\theta}{\theta}\right) = 0$$

and therefore $$\lim_{\theta\to+0}\frac{\sin\theta}{\theta} = 1$$

But $\dfrac{\sin\theta}{\theta}$ is an even function of θ

and hence we have our result

$$\lim_{\theta \to 0} \frac{\sin \theta}{\theta} = 1$$

Derivative of sin *x*, cos *x*, and tan *x*

Let $\quad y = \sin x$

then $\quad \frac{dy}{dx} = \lim_{h \to 0} \frac{\sin(x+h) - \sin x}{h}$

$$= \lim_{h \to 0} \frac{\sin x \cos h + \cos x \sin h - \sin x}{h}$$

Now $\lim_{h \to 0} \cos h = 1$ since $\cos x$ is a continuous function of x. Hence our limit reduces to

$$\lim_{h \to 0} \frac{\cos x \sin h}{h} = \cos x$$

whence

$$\frac{d(\sin x)}{dx} = \cos x \tag{2.16}$$

In the same way we may show that

$$\frac{d(\cos x)}{dx} = -\sin x \tag{2.17}$$

and

$$\frac{d(\tan x)}{dx} = \frac{d}{dx} \frac{\sin x}{\cos x} = \frac{\sin^2 + \cos^2 x}{\cos^2 x}$$

$$= \sec^2 x \tag{2.18}$$

For completeness we give the other derivatives leaving proof to the reader

$$\frac{d(\operatorname{cosec} x)}{dx} = -\operatorname{cosec} x \cot x \tag{2.19}$$

$$\frac{d(\sec x)}{dx} = \sec x \tan x \tag{2.20}$$

$$\frac{d(\cot x)}{dx} = -\operatorname{cosec}^2 x \tag{2.21}$$

The Inverse Trigonometric Functions

Let $y = \sin x$, then we define $x = \sin^{-1} y$ or arc sin y as the inverse sine function.

Similar definitions hold for the other trigonometric ratios. More generally, if

$$y = \sin^{-1}\frac{x}{a} \quad \text{then} \quad \frac{x}{a} = \sin y$$

We must now find the derivatives of these functions.

Let $$y = \sin^{-1}\frac{x}{a} \quad \text{so that} \quad \frac{x}{a} = \sin y$$

$$\frac{dy}{dx} = 1/\frac{dx}{dy} = \frac{1}{a\cos y} = \frac{1}{a\sqrt{1-x^2/a^2}}$$

i.e. $$\frac{d}{dx}\left(\sin^{-1}\frac{x}{a}\right) = \pm\frac{1}{\sqrt{(a^2-x^2)}} \tag{2.22}$$

where we take the positive sign if a is positive and the minus sign if a is negative.

By treating the other inverse functions in the same way, the reader can easily prove the results

$$\frac{d}{dx}\cos^{-1}\frac{x}{a} = \mp\frac{1}{\sqrt{(a^2-x^2)}} \tag{2.23}$$

$$\frac{d}{dx}\tan^{-1}\frac{x}{a} = \frac{1}{a^2+x^2} \tag{2.24}$$

$$\frac{d}{dx}\operatorname{cosec}^{-1}\frac{x}{a} = \mp\frac{1}{x\sqrt{x^2-a^2}} \tag{2.25}$$

$$\frac{d}{dx}\sec^{-1}\frac{x}{a} = \pm\frac{1}{x\sqrt{x^2-a^2}} \tag{2.26}$$

$$\frac{d}{dx}\cot^{-1}\frac{x}{a} = -\frac{1}{a^2+x^2} \tag{2.27}$$

With, of course, the special case of $a = 1$ as of interest.

Exponential and Logarithmic Functions

The student should study carefully Example 10 on p. 30.

Let $$\phi_n(x) = (1+x/n)^n$$

We assume x positive for the present.
Expanding as in the example referred to

$$\phi_n(x) = 1+x+\frac{x^2}{1.2}\left(1-\frac{1}{n}\right)+\frac{x^3}{1.2.3}\left(1-\frac{1}{n}\right)\left(1-\frac{2}{n}\right)\dots$$

now all terms are > 0 so $\phi(x)$ is an increasing function of n,

but $$\phi_n(x) < 1+x+\frac{x^2}{2!}+\frac{x^3}{3!}+\dots+\frac{x^n}{n!}$$

and hence $$< 1+x+x\left[\frac{x}{2}+\frac{x^2}{2^2}+\dots+\frac{x^{n-1}}{2^{n-1}}\right]$$

$$< 1+x+\frac{x^2}{2-x}$$

But clearly $$\phi_n(x) > 1+x$$

Let $$\lim_{n\to\infty}\phi_n(x) = \phi(x)$$

This limit exists because $\phi_n(x)$ is an increasing function of n but is bounded above and

$$\phi(x) > 1+x \quad \text{but} \quad \phi(x) < 1+x+\frac{x^2}{2-x}$$

Consider the limit

$$\lim_{x\to 0} \frac{\phi(x)-1}{x}$$

We have $\dfrac{\phi(x)-1}{x} > 1$ but $<1+\dfrac{x}{2-x}$

whence $\displaystyle\lim_{x\to 0} \frac{\phi(x)-1}{x} = 1$

Now we showed in our example (p. 31) that

$$\lim_{n\to\infty}\left(1+\frac{1}{n}\right)^n = e \quad \text{a const.}$$

Consider the function of x and n defined by

$$\lim_{n\to\infty}\left(1+\frac{x}{n}\right)^n \tag{2.28}$$

If we note that x is constant as far as the limiting process with n is concerned, if we replace n by Mx then, when $n \to \infty$ so does Mx, or, since x is constant in the limiting process then $M \to \infty$ with n.

So we may write (2.28) as

$$\lim_{M\to\infty}\left(1+\frac{x}{Mx}\right)^{Mx} = \lim_{M\to\infty}\left(1+\frac{1}{M}\right)^{Mx}$$

$$= \lim_{M\to\infty}\left[\left(1+\frac{1}{M}\right)^M\right]^x = e^x$$

Hence $\displaystyle\lim_{m\to\infty}\left(1+\frac{x}{m}\right)^m = e^x$

But this is the function $\phi(x)$ we have just discussed. We may now deduce the result

$$\lim_{x\to 0}\frac{e^x-1}{x} = 1$$

We have now defined the function e^x.

We may define the function $\log x$ where the base of the logarithm is e as follows,

Let $$y = e^x$$

then $$x = \log_e y$$

i.e. $\log x$ is the inverse function of e^x.

The Derivative of e^x

By the definition

$$\frac{d}{dx} e^x = \lim_{h \to 0} \frac{e^{x+h} - e^x}{h}$$

$$= \lim_{h \to 0} \frac{(e^h - 1)e^x}{h} = e^x$$

using our previous result.

Hence $\frac{d}{dx} e^x = e^x$ and so $\frac{d}{dx} e^{(ax+b)} = ae^{(ax+b)}$

The Derivative of $\log x$

$$\frac{d}{dx} \log x = \lim_{h \to 0} \frac{\log (x+h) - \log x}{h}$$

$$= \lim_{h \to 0} \left[\frac{1}{h} \log \left(1 + \frac{h}{x}\right)\right]$$

Putting $\frac{h}{x} = \frac{1}{m}$ (we assume $x \neq 0$)

then when $h \to 0$, $m \to \infty$ and the limit becomes

$$\lim_{m \to \infty} \left[\frac{m}{x} \log \left(1 + \frac{1}{m}\right)\right] = \frac{1}{x} \lim_{m \to \infty} \left[\log \left(1 + \frac{1}{m}\right)^m\right]$$

$$= \frac{1}{x} \log e = \frac{1}{x}$$

since by definition $\log e = 1.$

Whence the result

$$\frac{d}{dx}\log x = \frac{1}{x} \quad \text{and} \quad \frac{d}{dx}\log(ax+b) = \frac{a}{ax+b}$$

Hyperbolic Functions

These functions bear a certain resemblance to the trigonometric functions. There are six types, referred to as the hyperbolic sine, hyperbolic cosine, etc., and written sinh x, cosh x, etc. They are defined as follows:

$$\sinh x = \tfrac{1}{2}(e^x - e^{-x}) \qquad \operatorname{cosech} x = \frac{1}{\sinh x}$$

$$\cosh x = \tfrac{1}{2}(e^x + e^{-x}) \qquad \operatorname{sech} x = \frac{1}{\cosh x}$$

$$\tanh x = \frac{\sinh x}{\cosh x} \qquad \coth x = \frac{1}{\tanh x}$$

It is casy to provc thc following identities from the definition

$$\cosh^2 x - \sinh^2 x = 1$$
$$1 - \tanh^2 x = \operatorname{sech}^2 x$$
$$1 + \operatorname{cosech}^2 x = \coth^2 x$$

There are also addition formulae:

$$\sinh(x \pm y) = \sinh x \cosh y \pm \cosh x \sinh y$$
$$\cosh(x \pm y) = \cosh x \cosh y \pm \sinh x \sinh y$$

and so

$$\sinh 2x = 2 \sinh x \cosh x$$
$$\cosh 2x = \sinh^2 x + \cosh^2 x$$

Note that sinh x may take any value from $-\infty$ to $+\infty$ and that cosh x is always positive and $\geqq 1$.

Inverse Hyperbolic Functions

These are defined in the usual manner.

If $x = \sinh y$ then $y = \sinh^{-1} x$ — the inverse function. These inverse functions can be expressed as logarithms for

$$x = \tfrac{1}{2}(e^y - e^{-y})$$

or,
$$e^{2y} - 2xe^y - 1 = 0$$

so
$$e^y = x \pm \sqrt{x^2+1}$$

We need only consider the + sign since e^y must be > 0.

Hence
$$y = \sinh^{-1} x = \log (x + \sqrt{x^2+1})$$

We may similarly show that

$$\cosh^{-1} x = \log (x \pm \sqrt{x^2-1})$$

Here we must retain both signs for all values of $x > 1$.

In exactly the same manner we may show that

$$\tanh^{-1} x = \tfrac{1}{2} \log \frac{1+x}{1-x} \quad \text{where} \quad x^2 < 1$$

and
$$\coth^{-1} x = \tfrac{1}{2} \log \frac{x+1}{x-1} \quad \text{where} \quad x^2 > 1$$

The Derivatives of Hyperbolic Functions

These are readily found from the definition to be

$$\frac{d}{dx} \sinh x = \cosh x \qquad \frac{d}{dx} \cosh x = \sinh x$$

$$\frac{d}{dx} \tanh x = \text{sech}^2 x \qquad \frac{d}{dx} \coth x = -\textit{cosech}^2\, x$$

$$\frac{d}{dx} \text{cosech}\, x = -\text{cosech}\, x \coth x$$

$$\frac{d}{dx}\operatorname{sech} x = -\operatorname{sech} x \tanh x$$

Derivatives of Inverse Functions

Using the definitions we obtain

$$\frac{d}{dx}\sinh^{-1} x = \frac{1}{\sqrt{x^2+1}}$$

$$\frac{d}{dx}\cosh^{-1} x = \pm\frac{1}{\sqrt{x^2-1}}$$

$$\frac{d}{dx}\tanh^{-1} x = \frac{1}{1-x^2} \qquad x^2 < 1$$

$$\frac{d}{dx}\coth^{-1} x = \frac{-1}{x^2-1} \qquad x^2 > 1$$

Note that $\dfrac{d}{dx}\cosh^{-1} x$ has two values. We take the + sign where $\cosh^{-1} x$ is positive and vice versa.

As an exercise the student should write down the corresponding formulae for the derivatives of $\sinh \dfrac{x}{a}$, $\sinh^{-1} \dfrac{x}{a}$, etc.

Composite Algebraic and Trigonometrical Functions

Using the theorems for the derivative of a product and of the function of a function, it is usually quite easy to write down the derivative of a composite function.

Example. Find the derivative of the function

$$y = \sqrt{(\sin^2 x + x^2 \cos x)}$$

We can break this down by putting

$$u = \sin^2 x + x^2 \cos x$$

and, of course,

$$y = u^{\frac{1}{2}}$$

Then

$$\frac{dy}{dx} = \frac{dy}{du}\frac{du}{dx}$$

$$= \tfrac{1}{2}u^{-\frac{1}{2}}.\frac{d}{dx}(\sin^2 x + x^2 \cos x)$$

$$= \frac{1}{2\sqrt{(\sin^2 x + x^2 \cos x)}}(2 \sin x \cos x + 2x \cos x - x^2 \sin x)$$

$$= \frac{2 \cos x (\sin x + x) - x^2 \sin x}{2\sqrt{(\sin^2 x + x^2 \cos x)}}$$

Examples

Find the derivatives of:

1. $\sin^m x$
2. $\tan^n x$
3. $\sin (\cos x)$
4. $\tan^3 x \sin x$
5. $\sqrt{x} \sin \sqrt{x}$
6. $\log (\sin x)$
7. $\text{arc} \sin x^2$
8. $\text{arc} \tan \sqrt{x^2-4}$
9. $\log \left(\frac{a+\sqrt{x}}{a-\sqrt{x}}\right)$
10. $e^x (x-1)^3$

Repeated Differentiation

By applying the rules for differentiation to the derivative of $f(x)$, i.e. to the function $f'(x)$ we can obtain the second derivative of $f(x)$ which is commonly written $f''(x)$, i.e.

$$\frac{d}{dx} f'(x) = f''(x)$$

Another common notation for $f''(x)$ is

$$\frac{d^2}{dx^2}f(x)$$

It may be possible to continue this process and to arrive at the nth derivative of a function. When this is so we use the notation

$$f^{(n)}(x) \quad \text{or} \quad \frac{d^n}{dx^n}f(x)$$

In only a few cases is there a general formula for the nth derivative of a function.

An important case is the formula for the nth derivative of the product of two functions u and v. This is known as *Leibnitz's Theorem.*

$$\frac{d^n}{dx^n}(uv) = \frac{d^n u}{dx^n}.v+\binom{n}{1}\frac{d^{n-1}u}{dx^{n-1}}\frac{dv}{dx}+\binom{n}{2}\frac{d^{n-2}u}{dx^{n-2}}\frac{d^2v}{dx^n}$$

$$+\ldots+u\frac{d^n v}{dx^n} \qquad (2.28)$$

This theorem is most easily proved by the method of mathematical induction. For we can easily show that

$$\frac{d}{dx}(uv) = v\frac{du}{dx}+u\frac{dv}{dx}$$

$$\frac{d^2}{dx^2}(uv) = \frac{d^2u}{dx^2}.v+2\frac{du}{dx}\frac{dv}{dx}+\frac{d^2v}{dx^2}u$$

and so by continuing we observe that we are forming the well known binomial coefficients.

If we write $(uv)_n$ for $\frac{d^n}{dx^n}(uv)$ and $(u)_r$ for $\frac{d^r}{dx^r}(u)$ etc., and denote the binomial coefficients by the symbols $\binom{n}{r}$ where

$\binom{n}{r}$ is the binomial coefficient associated with the factor $(u)_{n-r}(v)_r$, then the proof follows easily.

For let us assume that the coefficient of $(u)_{n-r}(v)_r$ in the expression for $(uv)_n$ is $\binom{n}{r}$. Differentiate this expression for $(uv)_n$ to form $(uv)_{n+1}$. It will then be found that the coefficient of $(u)_{n+1-r}(v)_r$ is $\binom{n+1}{r}$. Applying the principle of induction this proves the theorem.

Examples

1. If $y = \sqrt{x^2-a^2}$ find $\dfrac{d^2y}{dx^2}$

2. If $y = \sin^2 x$ find $\dfrac{d^2y}{dx^2}$ and $\dfrac{d^3y}{dx^3}$

3. Find $\dfrac{d^ny}{dx^n}$ if $y = x\,e^x$

4. If $y = a\cos(nx-b)$ show that $\dfrac{d^2y}{dx^2}+n^2y = 0$

5. If $y = \tan^{-1}\dfrac{x}{a}$ find $\dfrac{dy}{dx}$ and $\dfrac{d^2y}{dx^2}$

The Significance of the Sign of $f'(x)$

If $f'(x)$ is positive at the point x, then $f(x)$ is increasing at the point x. Similarly if $f'(x)$ is negative at any point then $f(x)$ is decreasing at that point.

We can easily formalise this statement; for if $f'(x_0)$ is positive then $\lim\limits_{h\to 0}\dfrac{f(x_0+h)-f(x_0)}{h}$ converges to a positive limit when $h \to 0$, i.e. $f(x_0+h)-f(x_0)$ has the same sign as h

i.e. $f(x_0+h) > f(x_0)$ and $f(x_0-h) < f(x_0)$ (if $h > 0$)

The next theorem is due to Rolle.

Rolle's Theorem

Let $f(x)$ be a function which has a derivative at all values of x under consideration. Then if

$$f(a) = 0 \quad \text{and} \quad f(b) = 0$$

then there is at least one value of x lying between a and b for which $f'(x) = 0$.

Before giving a formal proof, it is useful to look at the geometrical significance which makes the theorem look almost obvious. It simply states that a continuous curve which cuts the x axis at $x = a$ and $x = b$ must have at least one point between a and b for which the tangent is parallel to the x axis.

In Figure 7 there are three points and we have drawn a curve for which the function is positive. The truth for the

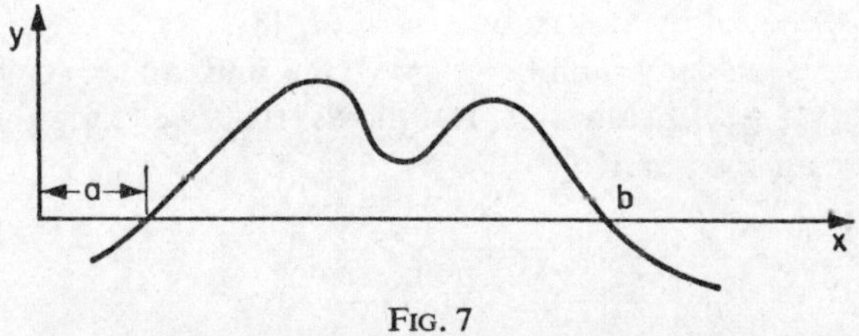

FIG. 7

case where the function is negative is easy to illustrate in the same way. The reader should at this stage be sufficiently indoctrinated to realise that proofs depending on geometrical notions are basically unsatisfying since the validity or otherwise of statements such as that in Rolle's Theorem have nothing to do with geometry.

We could in fact have equally well given an illustration for mechanics.

Let $f(t)$ represent the displacement at time t of a point which moves on a straight line: the distance of a motor car along a straight road for example.

The condition $f(t_0) = 0 = f(t_1)$ means that at time t_0 the car is at the starting point and that at time t_1 it has returned there. Rolle's Theorem states that the car must have stopped at least once in the intervening time. In the simplest case this would be once and it is obvious that this is where the driver reversed direction.

However valuable such illustrations may be, they are no substitute for a mathematical proof. We can supply one as follows:

Since $f'(x)$ exists, then $f(x)$ is continuous. It is, therefore, bounded. Hence there is a number K such that $f(x) \leq K$ for all values of x under consideration. Under these conditions the function $f(x)$ is said to have an upper bound M (and since the function is continuous there is at least one value of x, say $x = \xi$, for which $f(\xi) = M$).

No value of $f(x)$ can exceed M but any number less than M is exceeded by at least one value of $f(x)$.

For given any number δ we can find an x so that $f(x) > M-\delta$. From this it follows that we can always choose an x so that

$$\frac{1}{M-f(x)} > \frac{1}{\delta}$$

But this means that the function $1/[M-f(x)]$ is not continuous. But $M-f(x)$ is certainly continuous and thus $1/[M-f(x)]$ must also be continuous except where the denominator vanishes.

There must therefore be a point at which the denominator vanishes, and at this point $f(x) = M$. At this point which we have called $x = \xi$, $f'(\xi) = 0$. For if it were positive we could find a neighbouring point $x+\eta$ at which $f(x+\eta) > M$, and this is impossible. By similar arguments we can show that $f'(\xi)$ cannot be negative. This proves the theorem.

As an exercise the reader should show that the theorem holds good for a function $f(x)$ which is such that

$$f(a) = c = f(b)$$

where c is a constant.

Maxima and Minima

We say that the function $f(x)$ has a maximum at $x = \xi$ if the value of $f(\xi)$ is greater than the values of $f(x)$ at points near but on either side of $x = \xi$.

Expressing this formally, if we can find a number ε such that

$$f(\xi) > f(x)$$

for all points in the interval $\xi-\varepsilon < x < \xi$ and $\xi < x < \xi+\varepsilon$ then $f(x)$ has a maximum at the point $x = \xi$.

Similarly if the value $f(\xi)$ for $x = \xi$ is less than the values of $f(x)$ for points in the neighbourhood and on the other side of $x = \xi$, then $f(x)$ has a minimum at $x = \xi$.

Formally, if we can find a number ε so that

$$f(\xi) < f(x)$$

for all points in the interval $\xi = \varepsilon < x < \xi$ and $\xi < x < \xi+\varepsilon$ then $f(x)$ has a minimum at the point $x = \xi$.

Theorem. The necessary condition that $f(x)$ has a maximum or a minimum at $x = \xi$ is that $f'(\xi) = 0$.

This follows at once from the general discussion where we showed that if $f'(\xi) > 0$ there are values of x near ξ for which $f(\xi+h) > f(\xi)$. But by definition $f(\xi)$ is a maximum value of $f(x)$, hence $f'(\xi)$ cannot be positive.

Similarly it cannot be negative. Hence at a maximum value of $f(x)$,

$$f'(\xi) = 0$$

By similar arguments we may show that the condition is necessary for a minimum. Once again the apparently obvious nature of the theorem from geometrical considerations is illustrated by Figure 8.

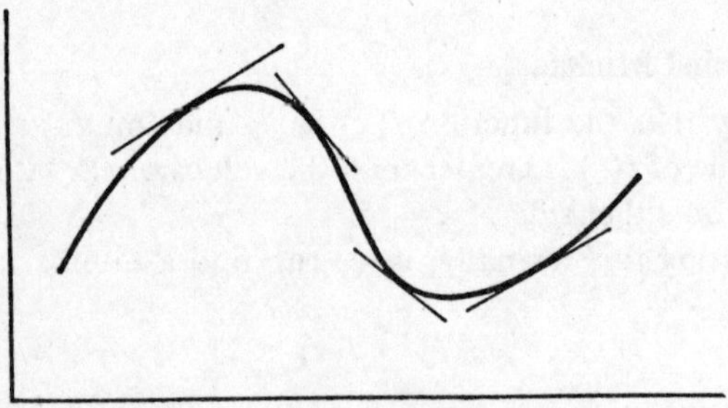

FIG. 8

The condition is not, however, sufficient, i.e. it does NOT follow that if $f'(x) = 0$ for $x = \xi$ that the function $f(x)$ has a maximum or minimum at $x = \xi$.

To illustrate this consider the function $y = x^3$. Here $dy/dx = 3x^2$ and hence at $x = 0$

$$\frac{dy}{dx} = 0$$

A glance at the graph of the function $y = x^3$ will show that there is neither a maximum nor a minimum at $x = 0$. It is, of course, possible to lay down conditions which will determine whether $f(x)$ has a maximum or a minimum at the value $x = \xi$.

If we study Figure 8 we can see that in the neighbourhood of a maximum, $f'(x)$, the slope of the tangent to the curve is in process of changing sign, and that as the function y passes through a maximum, the slope of the tangent changes from positive to negative. Similarly at a minimum, the slope of the tangent changes from negative to positive.

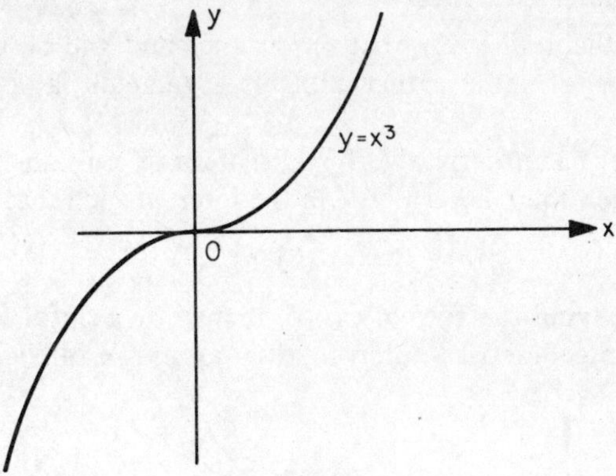

FIG. 9

Taking the first case this implies that $f'(x)$ is a decreasing function in the neighbourhood of the maximum at $x = \xi$, and similarly that $f'(x)$ is an increasing function in the neighbourhood of the minimum at $x = \xi$.

This gives a practical rule for determining whether we have a maximum or a minimum.

At a Maximum we have

1. $$f'(\xi) = 0$$
2. $$f''(\xi) < 0$$

At a Minimum we have

1. $$f'(\xi) = 0$$
2. $$f''(\xi) > 0$$

Note that the existence of $f'(x)$ does not of itself imply the existence of $f''(x)$ any more than the existence of $f(x)$ implies that of $f'(x)$, but in most practical cases it will be found that $f''(x)$ does exist.

The Mean Value Theorem

This theorem is of great importance and will be used as the basis of some rather difficult arguments later on. It states:

If $f(x)$ has a derivative for all values of x in an interval (a, b) then there is a value ξ in the interval such that

$$f(b)-f(a) = (b-a)f'(\xi) \tag{2.29}$$

Before giving the formal proof, it may be helpful to consider the geometrical interpretation as shown in Figure 10.

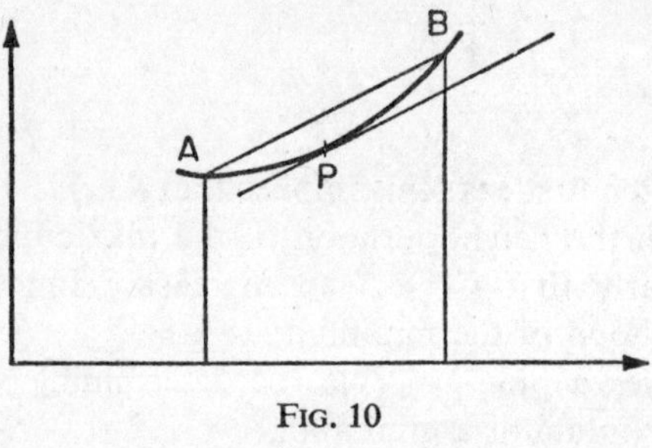

FIG. 10

The theorem merely states that there is a point P on the curve APB at which the tangent is parallel to the chord AB. The formal proof follows from Rolle's Theorem. Consider the function

$$y = f(b)-f(x)-\frac{b-x}{b-a}\{f(b)-f(a)\}$$

This clearly vanishes for $x = a$ and $x = b$, and hence there is some value ξ lying between a and b for which the derivative vanishes. The derivative is

$$\frac{f(b)-f(a)}{b-a}-f'(x)$$

hence there is a value ξ such that

$$f(b)-f(a) = (b-a)f'(\xi)$$

which proves the result.

It is sometimes convenient to replace ξ by $a+\theta(b-a)$ when $0 < \theta < 1$ and the theorem is written

$$f(b)-f(a) = (b-a)f'[a+\theta(b-a)] \tag{2.30}$$

or

$$f(x+h)-f(x) = hf'(x+\theta h) \tag{2.31}$$

An important deduction for the last theorem is that if $f'(x) = 0$ for all values of x in an interval, then $f(x)$ is constant in the interval.

For let a and b be two points in the interval, then

$$f(a)-f(b) = (b-a)f'[a+\theta(b-a)] = 0$$

$$f(a) = f(b) = \text{const.}$$

Examples

1. Using the Mean Value Theorem show that the value ξ in equation 2.29 takes the following forms:

   if $\quad y = x^2 \quad$ then $\quad \xi = \dfrac{a+b}{2}$

   if $\quad y = x^3 \quad$ then $\quad \xi = \sqrt{\tfrac{1}{3}(a^2+ab+b^2)}$

   Show that in the second case $a < \xi < b$.

2. If $y = \sin x$ use the theorem to show that under certain conditions it can be expressed in the form

$$\frac{\sin\theta}{\theta} = \cos\lambda\theta \qquad 0 < \lambda < 1$$

Hence deduce that $\lim_{\theta \to 0} \frac{\sin \theta}{\theta} = 1$

and show further that if powers of θ higher than θ^2 may be neglected then

$$\lambda = \frac{1}{\sqrt{3}}$$

3. Find the maximum and minimum values of the functions:

(a) $\frac{x}{1+x^2}$ (b) $x^3(1+x^2)$

4. Find the maximum values only of the functions:

(a) $x\,e^{-x}$ (b) $A\,e^{-kx^2}$

5. Find the minimum value of:

$a^2 \sec^2 x + b^2 \operatorname{cosec}^2 x.$

6. A cylinder is inscribed in a cone. Show that its volume is a maximum when its height is $\frac{1}{3}$ that of the cone.

CHAPTER THREE

DIFFERENTIAL CALCULUS

(Continued)

Differentials

When being introduced to the calculus, it is usual to impress on the student that in the expression for a derivative, dy/dx, the symbols must be taken as a whole and that dy and dx taken by themselves are meaningless.

It is now convenient to give a meaning to these symbols. There is a degree of subtlety about it which makes it necessary to postpone discussion until the elementary notion of differentiation has been fully grasped. Once more geometrical ideas make the idea easier to grasp in the first instance.

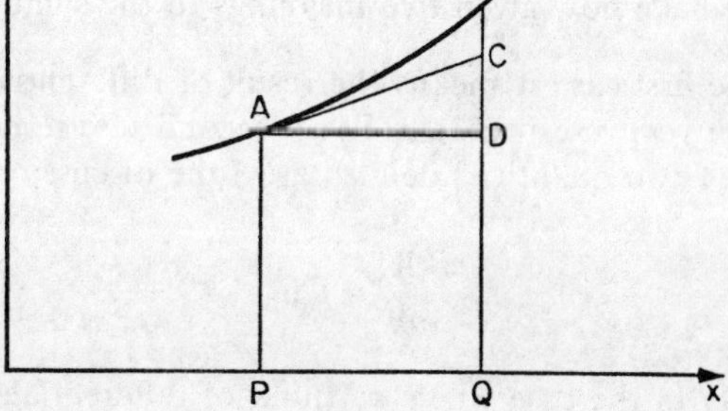

FIG. 11

Consider the portion of the curve $y = f(x)$, AB as in Figure 11. AC is the tangent at A and AD is parallel to the

x axis. Let x take an increment δz and let δy be the corresponding increment in y. So

$$\delta x = AD \quad \text{and} \quad \delta y = BD$$

CD is, of course, equal to $\delta x f'(x)$. We now define dy as

$$dy = f'(x)\delta x$$

In the special case $y = x$ this definition gives

$$dy = dx = \delta x$$

In fact δx and dx are the same. We therefore write

$$dy = f'(x)dx$$

or

$$\frac{dy}{dx} = f'(x)$$

Here, however, we mean by dy/dx the ratio of two quantities. It is convenient in most cases to think of dy and dx as being small but there is no need for this. It is their *ratio* which is important.

We hope that the reader will not be confused by the fact that we have now given two meanings to the symbol $\frac{dy}{dx}$.

In the first case it means the result of differentiating the function y with respect to x. In the second case it means the ration of two quantities defined as in the discussion above. In either case

$$\frac{dy}{dx} = f'(x)$$

but only in the case when we think of differentials can we write

$$dy = f'(x)\, dx$$

The use of differentials is important when we come to the

discussion of functions of two or more variables to which we must now proceed.

Functions of Two or More Variables

We must now consider the meaning and properties of a function which depends on two or more independent variables.

Consider first the function

$$z = f(x, y) \tag{3.1}$$

Since it is easy to visualise, let us imagine that z represents height above sea level of some point while x and y are the coordinates of position in the plane—grid reference for example.

If we assume that z is constant, $z = 1000$ ft for example, then the relation

$$1000 = f(x, y)$$

defines a curve. This is the contour line for the height of 1000 ft as shown on an Ordnance Survey map for example. Other contour lines can be represented by giving z different values.

If we choose too large a value of z there are no contour lines. This can be deduced from the equation, because in such cases there are no real values of x and y which satisfy it.

If we assume that the Y axis runs North–South and the X axis East–West, then by assuming a constant value of y, say y_0, we obtain

$$z = f(x, y_0)$$

This curve is that produced by the intersection of the surface with a plane running E–W through all points for which $y = y_0$.

Let us now take a mechanical example. Consider the displacement of an elastic string stretched between two fixed points which we may take to be $x = 0$ and $x = l$. Let the vertical displacement of a point on the string be $y = y(x)$ as represented in Figure 12.

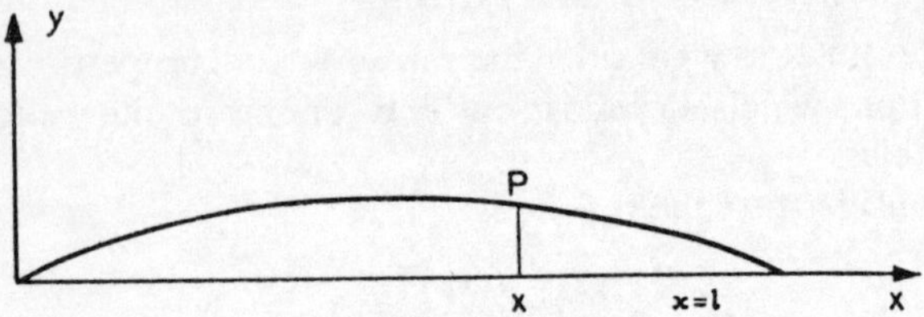

FIG. 12

The position of a point P on the string is a function of x but also of the time t if the string is free to vibrate. In other words

$$y = f(x, t) \tag{3.2}$$

If we consider t to be constant $= t_0$ then we are left with the relation

$$y = f(x, t_0) \tag{3.3}$$

which gives the shape to the string at time t_0.

On the other hand if we take x as fixed, say $x = x_0$, we have

$$y = f(x_0, t) \tag{3.4}$$

This gives the displacement of the point P for which $x = x_0$ as a function of the time.

Thus equation (3.2) gives us a complete picture of the behaviour of the string for all points of its length and for all times.

We can easily extend this to the displacement z of a surface such as an elastic membrane—a drum for example. Here there are three independent variables and we have

$$z = f(x, y, t) \tag{3.5}$$

If for example we put $t = t_0$ we have the equation

$$z = f(x, y, t_0)$$

which gives the shape of the displaced membrane at time $t = t_0$.

As a last example, this time from the science of meteorology, consider the temperature of the atmosphere at any point of space and time. To completely specify this temperature θ we need an equation

$$\theta = f(x, y, z, t) \tag{3.6}$$

Here there are four independent variables: position on the earth's surface, height, and time.

It is intuitively obvious that the expressions which we have been giving are usually valid only for some limited range of the variables, i.e. for a definite region of space. The equation for the membrane applies for the portion bounded by, e.g., the edges of the drum. The temperature of the earth's atmosphere is not intended to have relevance for z negative, etc.

It is now time to enquire what we mean by derivatives of functions of more than one independent variable.

Continuity of Functions of Two or More Variables

All the difficulties associated with the property of continuity—or the absence of it—which we have already studied in the case of functions of a single variable, apply in even greater measure in the case of functions of more than one variable.

In what follows we shall assume that we have to deal only with continuous functions.

We shall give a definition of continuity for a function of two variables. The extension to cases of more than two is easy.

Definition. A function $f(x, y)$ of two independent variables x and y is said to be continuous at the point a, b if

$$\lim_{\substack{h\to 0\\ k\to 0}} \{f(a+h, b+k)-f(a, b)\} = f(a, b) \tag{3.7}$$

where the statement is true no matter how h and k tend to zero.

Partial Differentiation

Let $z = f(x, y)$ be a continuous function of x and y. We define the partial derivatives of z as follows:

$$\frac{\partial z}{\partial x} = \frac{df}{dx}$$

where in performing the differentiation we regard y *as constant*. Similarly

$$\frac{\partial z}{\partial y} = \frac{df}{dy}$$

where now in performing the differentiation *we regard* x *as constant*. The 'curly' operators $\partial/\partial x$ and $\partial/\partial y$ signify this.

Example. Let $$z = x^2 \sin y$$

then, $$\frac{\partial z}{\partial x} = 2x \sin y \quad \text{and} \quad \frac{\partial z}{\partial y} = x^2 \cos y$$

The reader must not suppose that he is being confronted with some new or difficult idea. The notation tends to suggest this, but in fact, partial differentiation is carried out exactly as ordinary differentiation with the proviso that in doing it we consider all the variables except one as constant during the operation.

We can, of course give a formal definition.

Definition. If $f(x, y)$ is a function of the independent variables x and y and if the limits

$$\lim_{h\to 0}\frac{f(x+h,\, y)-f(x,\, y)}{h} \quad \text{and} \quad \lim_{k\to 0}\frac{f(x,\, y+k)-f(x,\, y)}{k}$$

both exist then they are the partial derivatives of $f(x, y)$ with respect to x and y and are written respectively

$$\frac{\partial f}{\partial x} \quad \text{and} \quad \frac{\partial f}{\partial y}$$

Another notation frequently used is

$$f'_x \text{ for } \frac{\partial f}{\partial x} \quad \text{and} \quad f'_y \text{ for } \frac{\partial f}{\partial y}$$

Note that the idea that x and y are independent is fundamental. Clearly if y depended on x we could not increase x to $x+h$ without affecting y and vice-versa.

Cases do occur where we find it a matter of practical convenience to use the symbolism of partial differentiation even if x and y are not independent. Suppose that both x and y are given as functions of a third variable t

$$x = \phi(t) \quad \text{and} \quad y = \psi(t)$$

then if we have a function given in the form

$$z = f(x, y) = f\{\phi(t),\, \psi(t)\} \tag{3.8}$$

Here z is a function of the single variable t. We may wish to find the value of dz/dt. In this case the symbolism of partial differentiation is helpful. We shall explain this in a moment.

Extension to Three or More Variables

If, as before,

$$\theta = f(x, y, z, t)$$

then $\partial\theta/\partial t$ means df/dt where in performing the differentiation we regard x, y, and z as constants. Similar meanings attach to $\partial\theta/\partial x$, $\partial\theta/\partial y$, etc.

Clearly $\partial\theta/\partial t$ means the rate of change of temperature at a fixed point in space and $\partial\theta/\partial z$, $\theta\partial/\partial x$, $\partial\theta/\partial y$ give the temperature gradients, as they are called, at a fixed time in the vertical and horizontal directions.

The Total Derivative

We return to the problem of finding dz/dt when z is a function of x and y but when both x and y are functions of a single variable t. Let

$$z = f[\phi(t), \psi(t)] \quad \text{i.e.} \quad z = F(t)$$

We wish to find dF/dt.

We could of course try to find $F(t)$ as an explicit function of t and use the ordinary methods. It is frequently simpler to proceed as follows:

Let t increase to $t+\tau$ and hence x to $x+h$ and y to $y+k$. By the definition of a derivative

$$F'(t) = \lim_{\tau\to 0} \frac{1}{\tau}[f\{\phi(t+\tau), \psi(t+\tau)\} - f\{\phi(t), \psi(t)\}]$$

$$= \lim_{\tau\to 0}\left[\frac{f(x+h, y+k)-f(x, y+k)}{h}\cdot\frac{h}{\tau}+\frac{f(x, y+k)-f(x, y)}{k}\cdot\frac{k}{\tau}\right]$$

Now apply the Mean Value Theorem. This gives

$$f(x+h, y+k)-f(x, y+k) = hf'_x(x+\theta h, y+k)$$

where we write f'_x instead of f' because there has been no change in y during the operation and it has been treated as though it were a constant. Similarly

$$f(x, y+k)-f(x, y) = kf'_y(x, y+\theta' k)$$

Now by definition, $\lim \dfrac{h}{\tau} = \phi'(t)$ and $\lim \dfrac{h}{\tau} = \psi'(t)$

and also

$$\lim_{\tau\to 0} f_x'(x+\theta h, y+k) = f_x'(x, y) \;\&\; \lim_{\tau\to 0} f_y'(x, y+\theta' h) = f_y'(x, y)$$

so equation reduces to

$$F'(t) = f_x'(x, y) . \psi'(t) + f_y'(x, y)\psi'(t)$$

when τ tends to zero. This expression is often written in the equivalent form

$$\frac{df}{dt} = \frac{\partial f}{\partial x}\frac{dx}{dt} + \frac{\partial f}{\partial y}\frac{dy}{dt} \qquad (3.9)$$

These formulae are of great importance.

Note that in carrying out the partial differentiations we proceed as though x and y were independent. *After* we have done this, and *only after*, we may substitute their values in terms of t and so find $F(t)$ in terms of t.

df/dt is called the *Total Derivative* of $f(x, y)$ when x and y are related.

As a special case consider the replacement of t by x. We then have

$$\frac{df}{dx} = \frac{\partial f}{\partial x} + \frac{\partial f}{\partial y}\frac{dy}{dx} \qquad (3.10)$$

The Mean Value Theorem for Functions of Two Variables

Let $f(x, y)$ be a function of the independent variables x and y and let $\frac{\partial f}{\partial x}$ and $\frac{\partial f}{\partial y}$ exist and be continuous, then

$$\begin{aligned} f(x+h, y+k) &= f(x, y) + hf_x'(x+\theta h, y+\theta k) \\ &\quad + kf_y'(x+\theta h, y+\theta k), \quad [0 < \theta < 1] \end{aligned} \qquad (3.11)$$

We shall not give a formal proof of this theorem but merely state it.

However, if f_x' and f_y' are continuous

$$f_x'(x+\theta h, y+\theta k) = f_x'(x, y) + \varepsilon_1(h, k) \qquad (3.12)$$

and $$f'_y(x+\theta h, y+\theta k) = f'_y(x, y)+\varepsilon_2(h, k) \tag{3.13}$$

where ε_1 and ε_2 tend to zero when h and k tend to zero.

If we replace h and k by δx and δy then (3.11) may be expressed

$$\delta z = f'_x(x, y)\delta x+f'_y(x, y)\delta y+\varepsilon_1\delta x+\varepsilon_2\delta y$$

where ε_1 and ε_2 are small when δx and δy are small.

We now *define* the differential dz by

$$dz = \frac{\partial f}{\partial x}\delta x+\frac{\partial f}{\partial y}\delta y$$

In the special cases $z = x$ and $z = y$

$$\delta z = dx \quad \text{and} \quad \delta y = dy$$

and so we can replace δx and δy by dx and dy and we then have the result

$$dz = \frac{\partial f}{\partial x}dx+\frac{\partial f}{\partial y}dy \tag{3.14}$$

dz is called a *Total Differential.*

We have already proved that if x and y are functions of t

$$\frac{dz}{dt} = \frac{\partial z}{\partial x}\frac{dx}{dt}+\frac{\partial z}{\partial y}\frac{dy}{dt} \tag{3.9}$$

If we now use the notion of differentials for functions of one variable we may regard dz/dt in the above formula as a ratio and multiply by dt

This gives us at once $\quad dz = \dfrac{\partial z}{\partial x}dx+\dfrac{\partial z}{\partial y}dy$

which is identical with (3.14)

The useful thing here is that it shows that the formula for a total differential applies whether x and y are independent or not.

We can extend the above ideas to functions of more than two variables and we should find for example that if

$$u = u(x, y, z, t)$$

$$du = \frac{\partial u}{\partial x}\,dx + \frac{\partial u}{\partial y}\,dy + \frac{\partial u}{\partial z}\,dz + \frac{\partial u}{\partial t}\,dt \qquad (3.15)$$

Partial Derivatives of Higher Order

By applying the methods of partial differentiation to the functions which are themselves in general functions of x and y we can form higher derivatives such as

$$\frac{\partial}{\partial y}\left(\frac{\partial f}{\partial x}\right) \quad \text{and} \quad \frac{\partial}{\partial y}\left(\frac{\partial f}{\partial y}\right) \quad \text{and} \quad \frac{\partial}{\partial x}\left(\frac{\partial f}{\partial x}\right)$$

these are written

$$\frac{\partial^2 f}{\partial x \partial y}, \quad \frac{\partial^2 f}{\partial y^2}, \quad \frac{\partial^2 f}{\partial x^2} \quad \text{or} \quad f'_{xy}, \quad f'_{yy}, \quad f'_{xx}$$

For example the expression $\dfrac{\partial^5 f(x, y)}{\partial x^2 \partial y^3}$ means that we have to differentiate $f(x, y)$ five times. First three times with respect to y and then twice with respect to x.

The Commutative Law for Partial Derivatives

This states that if all the functions are continuous

$$\frac{\partial^2 f(x, y)}{\partial x \partial y} = \frac{\partial^2 f(x, y)}{\partial y \partial x} \qquad (3.16)$$

In other words, it does not matter whether we differentiate first with respect to x and then with respect to y, or vice-versa. We shall not give a proof of this, but the student may find plenty of examples to illustrate the theorem.

Provided all functions are continuous, the theorem can be extended

e.g. $$\frac{\partial^{p+q+r}}{\partial x^p\,\partial y^q\,\partial z^r} = \frac{\partial^{p+q+r}}{\partial x^p\,\partial z^r\,\partial y^q} = \frac{\partial^{p+q+r}}{\partial z^r\,\partial x^p\,\partial y^q} = \ldots$$

Change of Variable

Let $$z = f(x, y)$$

and suppose that both x and y are themselves functions of two other independent variables u and v, so

$$x = x(u, v) \quad \text{and} \quad y = y(u, v)$$

then of course $$z = z(u, v)$$

We wish now to try to express $\frac{\partial z}{\partial u}$ and $\frac{\partial z}{\partial v}$ in terms of u and v. We already know that

$$dz = \frac{\partial z}{\partial x}\,dx + \frac{\partial z}{\partial y}\,dy$$

but $$dx = \frac{\partial x}{\partial u}\,du + \frac{\partial x}{\partial v}\,dv$$

and $$dy = \frac{\partial y}{\partial u}\,du + \frac{\partial y}{\partial v}\,dv$$

Substituting,

$$dz = \frac{\partial z}{\partial x}\left[\frac{\partial x}{\partial u}\,du + \frac{\partial x}{\partial v}\,dv\right] + \frac{\partial z}{\partial y}\left[\frac{\partial y}{\partial u}\,du + \frac{\partial y}{\partial v}\,dv\right]$$

$$= \left(\frac{\partial z}{\partial x}\frac{\partial x}{\partial u} + \frac{\partial z}{\partial y}\frac{\partial y}{\partial u}\right)du + \left(\frac{\partial z}{\partial x}\frac{\partial x}{\partial v} + \frac{\partial z}{\partial y}\frac{\partial y}{\partial v}\right)dv$$

But we know that

$$dz = \frac{\partial z}{\partial u}\,du + \frac{\partial z}{\partial v}\,dv$$

so we deduce the results

$$\frac{\partial z}{\partial u} = \frac{\partial z}{\partial x}\frac{\partial x}{\partial u} + \frac{\partial z}{\partial y}\frac{\partial y}{\partial u} \tag{3.17}$$

and

$$\frac{\partial z}{\partial v} = \frac{\partial z}{\partial x}\frac{\partial x}{\partial v} + \frac{\partial z}{\partial y}\frac{\partial y}{\partial v} \tag{3.18}$$

Example. Let $z = f(x, y)$

and suppose $x = r\cos\theta$ and $y = r\sin\theta$

Then using the transformation $r^2 = x^2 + y^2$ and $\tan\theta = y/x$ we may quickly obtain the formulae

$$\frac{\partial \theta}{\partial x} = -\frac{\sin\theta}{r}, \qquad \frac{\partial \theta}{\partial y} = \frac{\cos\theta}{r},$$

$$\frac{\partial r}{\partial x} = \cos\theta, \qquad \frac{\partial r}{\partial y} = \sin\theta$$

We may now apply this to get

$$\frac{\partial z}{\partial x} = -\frac{\sin\theta}{r}\frac{\partial z}{\partial \theta} + \cos\theta\,\frac{\partial z}{\partial r}$$

$$\frac{\partial z}{\partial y} = \frac{\cos\theta}{r}\frac{\partial z}{\partial \theta} + \sin\theta\,\frac{\partial z}{\partial r}$$

The student should proceed a stage further and prove the important result

$$\frac{\partial^2 z}{\partial x^2} + \frac{\partial^2 z}{\partial y^2} = \frac{\partial^2 z}{\partial r^2} + \frac{1}{r}\frac{\partial z}{\partial r} + \frac{1}{r^2}\frac{\partial^2 z}{\partial \theta^2} \tag{3.19}$$

The equation

$$\frac{\partial^2 z}{\partial x^2} + \frac{\partial^2 z}{\partial y^2} = 0 \tag{3.20}$$

is of great importance in physics. It is a special form of Laplace's Equation.

We shall see later, when we come to consider differential equations, that the solution can be made much easier to obtain if we choose a system of coordinates which take simple form on the boundaries. For example, if one of the boundaries is a circle we should obviously try to solve the equation in terms of polar coordinates, for the boundary becomes r = const.

In cases where there is circular symmetry so that the solution is independent of θ the equation is reduced to

$$\frac{\partial^2 z}{\partial r^2}+\frac{1}{r}\frac{\partial z}{\partial r} = 0 \tag{3.21}$$

which is now an 'ordinary' differential equation (see p. 165).

An Example in the Use of Differentials

We shall illustrate the use of differentials by an example from the science of meteorology.

A gas such as air has two specific heats. One, denoted by C_p, is the quantity of heat, in Joules, required to raise the temperature of 1 kilogram of air through 1°C, the pressure being kept constant. The other, C_v, is similarly defined for constant volume.

Values are $C_p = 1007$, $C_v = 718{\cdot}9$ and the ratio C_p/C_v, usually denoted by γ, is $1{\cdot}401$.

Now suppose we supply an amount of heat dq to a quantity of gas having volume v and at pressure p. This will result in a rise in temperature and also in work being done, if the gas is allowed to expand. In such a case the work done against the external pressure is $Apdv$ and the increase in thermal energy is $C_v dT$, so we have

$$dq = C_v dT + Apdv \tag{3.22}$$

where T is the absolute temperature and A is the reciprocal of the mechanical equivalent of heat. Now Boyle's Law states that

$$pv = RT \quad \text{where } R \text{ is a const.}$$

$$\therefore \quad p\,dv+v\,dp = R\,dT \tag{3.23}$$

$$\therefore \quad dq = (C_v+AR)\,dT-Av\,dp \tag{3.24}$$

If we could arrange for the pressure to be kept constant so that $dp = 0$ we would have

$$dq = C_p\,dT = (C_v+AR)\,dT \tag{3.25}$$

and we may deduce

$$C_p-C_v = AR$$

or

$$dq = C_p\,dT-Av\,dp$$

If there is no external heat supply and we use Boyle's Law

$$C_p\,dT = \frac{ART}{p}\,dp \tag{3.26}$$

so

$$C_p\frac{dT}{T} = (C_p-C_v)\frac{dp}{p} \quad \text{or} \quad \gamma\frac{dT}{T} = \gamma-1\frac{dp}{p}$$

Integrating this

$$\frac{T^\gamma}{p^{\gamma-1}} = \text{const.} \tag{3.27}$$

This is the Adiabatic Relation.

Using the formulae already obtained we can transfer this into the form

$$pv^\gamma = \text{const.} \tag{3.28}$$

We can now study the change in a rising parcel of air in the atmosphere

$$dp = -g\rho\,dz$$

where ρ is the density, z the height, and T the temperature at some point in the atmosphere.

Under adiabatic conditions, for a rising particle of air at temperature T'

$$\frac{dT'}{T'} = \frac{\gamma-1}{\gamma}\frac{dp}{p} \quad \text{or} \quad \frac{dT'}{dz} = -\frac{\gamma-1}{\gamma} g\rho \frac{T'}{R\rho T}$$

This gives the change in temperature of the rising particle in terms of that of the surrounding air which controls dp. In practice T and T' are nearly equal and we get

$$\frac{dT'}{dz} = -\frac{gA}{cp} \tag{3.29}$$

which is a constant. The value is about 9°C per km.

The point of interest here is that if the natural distribution of temperature is more rapid than this, rising air becomes relatively lighter than its surroundings as it ascends.

This is an unstable situation and fundamentally gives rise to the violent ascending air currents found in thunder clouds; although in such cases there can be release of heat by the condensation of water vapour into solid drops of rain.

Examples

Find the partial derivatives with respect to x and y for the following functions:

1. $e^x \sin y$
2. $\log (x^3+y^3)$
3. $\tan^{-1} \frac{x}{y}$
4. $\frac{e^{xy}}{x+y}$
5. $\log \sin (x-y)$
6. $x \log \tan (x+y)$
7. If $U = f(x-ct) + g(x+ct)$

   show that $$\frac{\partial^2 U}{\partial t^2} = c^2 \frac{\partial^2 U}{\partial x^2}$$

8. If $V = \log r$ where $r = (x-a)^2+(y-b)^2$

   show that $$\frac{\partial^2 V}{\partial x^2} + \frac{\partial^2 V}{\partial y^2} = 0$$

CHAPTER FOUR

INTEGRATION—REVISION

Introduction

Before getting down to the problem of integration, we think it may be worth while to digress a little and make a few general remarks on the correct attitude of mind to adopt in trying to solve mathematical problems.

So far no very serious difficulties have been encountered. We have been able to give more or less definite rules for the guidance of the student. The situation is analogous to making a journey along a road. Provided that we stick to the road and follow the sign posts, we can hardly go wrong. If we digress and take a wrong turning we shall almost certainly have to retrace our steps and come back to where we left the road. Short cuts are fraught with danger, although they may, with luck, sometimes shorten our route in a spectacular manner.

It goes without saying that we must be suitably equipped or we may fail to complete our journey—if boots disintegrate it is extremely difficult to complete a journey in bare feet.

Much of mathematics is like this. Provided we stick to the rules, avoid digressions, beware of short cuts, and are properly equipped with a sound grasp of algebra, elementary trigonometry and geometry, we should be able to tackle any problem in the differential calculus without undue difficulty.

When we pass on to the integral calculus, we enter a much more difficult field. Integration is a tentative process, and while every continuous function has an integral, this is very far from saying that we can give rules for finding it. There *are* some rules, of course, but in spite of them most students find integration very difficult, and not a few abandon further mathematical study because of this.

Let us revert to our analogy of a journey.

Suppose instead of travelling along a road, we wish to climb a mountain. Not, of course, a mountain in the Himalayas which has never been climbed before, but one which we know has been climbed many times in the past, but which to us is something new—some mountain in Scotland or Wales for example.

We ask ourselves what is the best method to set about this. There are no sign posts and the going is bound to be tough compared with walking along a road.

If we proceed step by step, with our eyes fixed on the few yards of ground immediately in front of us, we may eventually reach the summit, but it is likely to take an intolerably long time, and it is far more probable that we shall end by wandering around in some kind of circle. If, however, we are a little more experienced, the first thing we do is to stop and study the mountain from a distance. We can see the top, and we study the contours which lie between us and our goal, noting any serious snags such as a cliff face, a stream which has to be crossed, etc. After a few minutes study we can usually see how to get to the top, barring unforeseen obstacles and accidents. We may now proceed to the task of getting there. Now and only now has the time come to devote serious attention to the terrain immediately in front of us. There may well be some difficult patches and problems to be solved, but we know that as each is successively overcome we can be sure that we are making progress towards our goal on the summit.

Needless to say, what we said before about adequate equipment applies in even greater measure when we go mountaineering. We need even better equipment and more of it.

This analogy applies to a great many mathematical problems.

To proceed step by step, not really seeing where we are making for, is usually fruitless and exhausting. We should do better to stand back and look at the problem, see broadly how it can be solved and then set about filling the detailed steps to obtain the solution.

Ideally we should be in no doubt as to our ability to find the answer, but we cannot be sure at this stage of how much detail is required and how difficult individual steps may be.

If the student can approach the problem of integration with this philosophy, we think that many of the difficulties will disappear.

Having said this, the fact remains that the going is often rough, i.e. the algebra in the intermediate steps can often be complicated and tedious. It is, however, now time to leave this digression and get on with the revision of integration.

Definitions and General Remarks

The basic problem of integration is, of course, that of finding a function when we are given its derivative.

When we come to discuss the theory of differential equations we shall find that the solution usually can be reduced to that of evaluating an integral. Put differently, while we may be able in many cases to express the laws of nature in terms of differential equations, in order to find out what actually happens in a particular case we have to be able to find integral solutions.

Integration is therefore of great importance in the solution of physical problems.

To express these ideas formally we may say that the problem of integrating a function $f(x)$ is that of finding a function $\phi(x)$ such that

$$\frac{d}{dx}\phi(x) = f(x)$$

This is usually written as

$$\phi(x) = \int f(x)\,dx$$

It is reasonable to ask whether the function $\phi(x)$ always exists and if it is unique.

The answer is that if $f(x)$ is continuous then $\phi(x)$ always exists and is unique except for an arbitrary constant.

It does not follow that we can express the integral in simple terms. In fact, in general, we cannot do so. We can, however, develop rules based on a number of known results and these enable a wide variety of practically interesting problems to be solved.

Certain of these, listed below, are considered as standard formulae.

1. $\displaystyle\int x^n\,dx = \frac{x^{n+1}}{n+1}$ (if $n \neq -1$)
2. $\displaystyle\int \frac{dx}{x} = \log_e x$ ($n = -1$)
3. $\int \sin x\,dx = -\cos x$
4. $\int \cos x\,dx = \sin x$
5. $\int \sec^2 x\,dx = \tan x$
6. $\int \operatorname{cosec}^2 x\,dx = -\cot x$
7. $\displaystyle\int \frac{dx}{\sqrt{a^2-x^2}} = \sin^{-1}\left(\frac{x}{a}\right)$ or $\displaystyle -\cos^{-1}\left(\frac{x}{a}\right)$
8. $\displaystyle\int \frac{dx}{a^2+x^2} = \frac{1}{a}\tan^{-1}\left(\frac{x}{a}\right)$ or $\displaystyle -\frac{1}{a}\cot^{-1}\left(\frac{x}{a}\right)$
9. $\displaystyle\int \frac{dx}{x^2-a^2} = \tfrac{1}{2}a \log_e \frac{x-a}{x+a}$ if $x > a$

   $\displaystyle = \tfrac{1}{2}a \log_e \frac{a-x}{a+x}$ if $x < a$
10. $\int e^x\,dx = e^x$
11. $\int \sinh x\,dx = \cosh x$
12. $\int \cosh x\,dx = \sinh x$
13. $\int \operatorname{sech}^2 x\,dx = \tanh x$

14. $$\int \frac{dx}{\sqrt{x^2-a^2}} = \cosh^{-1}\frac{x}{a} = \log\{x+\sqrt{x^2-a^2}\}$$

15. $$\int \frac{dx}{\sqrt{x^2+a^2}} = \sinh^{-1}\frac{x}{a} = \log\{x+\sqrt{x^2+a^2}\}$$

16. $$\int \sqrt{a^2-x^2}\,dx = \tfrac{1}{2}x\sqrt{a^2-x^2}+\tfrac{1}{2}a^2\sin^{-1}\frac{x}{a}$$

17. $\int \sqrt{a^2+x^2}\,dx = \tfrac{1}{2}x\sqrt{a^2+x^2}+\tfrac{1}{2}a^2\log\{x+\sqrt{a^2+x^2}\}$

18. $\int \sqrt{x^2-a^2}\,dx = \tfrac{1}{2}x\sqrt{x^2-a^2}-\tfrac{1}{2}a^2\log\{x+\sqrt{x^2-a^2}\}$

Basic Types of Integrals for Algebraic Functions

The algebraic functions with which we are concerned are, as formerly, polynomials, rational functions, and algebraic functions of which the most general will be of the form $F(x, \sqrt{X})$ where

$$X = ax^2+2bx+c \quad \text{and} \quad a, b, c \text{ are constants}$$

We shall show later that the problem here can be reduced to the integration of five main types.

For the moment let us state the types with which we are concerned. These are:

Type I $\quad \displaystyle\int \frac{(Cx+D)\,dx}{ax^2+2bx+c}$ and the case where $C = 0, D = 1$

Type Ia $\quad \displaystyle\int \frac{dx}{ax^2+2bx+c}$

Type II $\quad \displaystyle\int \frac{(Cx+D)\,dx}{(ax^2+2bx+c)^n}$ where n is an integer

Type IIa $\quad \displaystyle\int \frac{dx}{(ax^2+2bx+c)^n}$ where $C = 0$ and $D = 1$

Type IIa can be reduced to Type Ia by means of what is called a reduction formula. This technique will be described shortly.

Type III $\int \frac{(Cx+D)\,dx}{\sqrt{(ax^2+2bx+c)}}$ with the special case

Type IIIa $\int \frac{dx}{\sqrt{(ax^2+2bx+c)}}$ and the related

Type IIIb $\int \sqrt{ax^2+2bx+C}\,dx$

Connected with the above is also

Type IV $\int \frac{dx}{(x-\alpha)\sqrt{ax^2+2bx+c}}$

This reduces to Type IIIa by a simple substitution

Type IVa $\int \frac{x^m\,dx}{\sqrt{(ax^2+2bx+c)}}$

This can be reduced to Type III.

Type IVb $\int \frac{dx}{(x-\alpha)^m\sqrt{(ax^2+2bx+c)}}$

This also can be reduced, to Type IV.

Finally we have

Type V $\frac{(Ux+V)\,dx}{(Ax^2+2Bx+C)\sqrt{(ax^2+2bx+c)}}$

This can be reduced, to Type I.

Techniques for Integration

We must now describe three techniques which are of great value in calculating integrals. The first of these is *Substitution, or Change, of Variable.*

Suppose we wish to integrate $\int f(x)\,dx$. It often happens that the problem can be made much easier if we introduce a new variable t so that $x = x(t)$ and then try to find the integral, such that

$$\int f[x(t)]\frac{dx}{dt}\,dt = \int F(t)\,dt$$

We can always express the result in terms of x by resubstitution.

Strange as it may appear, from the general formula above, the integral in terms of t is frequently much easier to find. The problem now is to think of the best substitution. Unfortunately we cannot give rules for every case which occurs. With experience, the student will be able to spot the correct substitution in many cases. The object is to try to reduce the integral in question to one of the Standard Forms.

One elementary approach with this end in view, is to make the trivial substitution which reduces a quadratic form such as $ax^2+2bx+c$ to the sum or difference of two squares, since it is thus that we give the formulae for Standard Forms. This is usually done mentally. When, as frequently happens, we have to deal with square roots, the obvious thing is to try to find a substitution which will rationalise the expression.

This is an important case and can be described as *Rationalisation* by *Substitution.*

It will be obvious that we can transform a rational function, say $R(\sqrt{x})$, into a rational function of t by the substitution $x = t^2$, for the function becomes

$$\int R(\sqrt{x})\,dx = \int R(t)2t\,dt$$

i.e. we have now to integrate a function of t which is rational.

It happens much more often that we have a quadratic

function of x under the square root. It must be fairly obvious that if we make the substitution

$$\sqrt{(ax^2+2bx+c)} = -t-x\sqrt{a}$$

then $$2bx+c = t^2-2\sqrt{a}.x.t$$

or $$x = \frac{t^2-c}{2(b+\sqrt{a}t)}$$

and $$\frac{dx}{dt} = \frac{\sqrt{a}(t^2+c)+2bt}{2(b+\sqrt{a}t)^2}$$

and $$t-x\sqrt{a} = \frac{\sqrt{a}(t^2+c)+2bt}{2(b+\sqrt{a}t)}$$

This has eliminated the square root and we have to deal with a rational function of t.

It is worth while to mention another very useful substitution which is of frequent application.

Expressions of the type

$$\int \frac{dx}{x^m\sqrt{ax^2+2bx+c}}$$

can be reduced by the substitution

$$x = \frac{1}{t} \quad \text{and so} \quad \frac{dx}{x} = -\frac{dt}{t}$$

to the form $$-\int \frac{t^m\,dt}{\sqrt{ct^2+2bt+a}}$$

The second technique is called Integration by Parts. It is based on the formula for differentiating a product.

$$\frac{d}{dx}(uv) = u\frac{dv}{dx}+v\frac{du}{dx}$$

$$\therefore \quad \int u \frac{dv}{dx}\, dx = uv - \int v \frac{du}{dx}\, dx$$

or $$\int u\, dv = uv - \int v\, du$$

or $$\int F(x)f'(x)\, dx = F(x)f(x) - \int f(x)F'(x)\, dx$$

If we write u_1 for $\int u\, dx$ we can express the formula in another way

$$\int (uv)\, dx = u_1 v - \int (u_1 v')\, dx$$

(To prove this it is only necessary to differentiate the product $u_1 v$.) It may well happen that $\int (u_1 v')\, dx$ is easier to evaluate than $\int (uv)\, dx$.

Occasionally it is useful to consider a simple function as a product of itself with unity,

e.g. $$\int \log x\, dx = x \log x - \int \frac{x}{x}\, dx = x \log x - x$$

or again

$$\int \sqrt{(a^2 - x^2)}\, dx$$

$$= x\sqrt{(a^2 - x^2)} + \int \frac{x^2\, dx}{\sqrt{(a^2 - x^2)}}$$

$$= x\sqrt{(a^2 - x^2)} - \int \sqrt{(a^2 - x^2)}\, dx + a^2 \int \frac{dx}{\sqrt{(a^2 - x^2)}}$$

Rearranging terms

$$\int \sqrt{(a^2 - x^2)}\, dx = \frac{x}{2}\sqrt{a^2 - x^2} + \frac{a^2}{2} \sin^{-1} \frac{x}{a}$$

The third technique is called *Successive Reduction*. We try to find a Reduction Formula which makes the problem simpler. It is easiest to illustrate the principle by an example.

Let $$u_n = \int x^n e^x\, dx$$

Integrate by parts

$$u_n = x^n e^x - n \int x^{n-1} e^x$$

$$= x^n e^x - nu_{n-1}$$

We have now reduced the problem to finding u_{n-1}. Clearly if we proceed in this manner, we can, if n is a positive integer, reduce the problem to finding u_0 or $\int e^x dx$, which is well known. If n is not an integer, but is positive, we can at least reduce the problem to its simplest form.

Consider a second example.

Let $$u_n = \int \sin^n x \, dx = \int \sin x \sin^{n-1} x \, dx$$

Integrating by parts

$$u_n = -\cos x \sin^{n-1} x + \int (n-1) \sin^{n-2} x \cos^2 x \, dx$$

$$= -\cos x \sin^{n-1} x + (n-1) \int \sin^{n-2} x \, dx - (n-1) \int \sin^n x \, dx$$

$$\therefore \quad u_n = \frac{-\cos x \sin^{n-1} x}{n} + \frac{n-1}{n} u_{n-2}$$

We have thus reduced the problem of finding u_n to that of finding u_{n-2}. Assuming n to be a positive integer, the problem reduces either to

$$\int \sin x \, dx \quad \text{or} \quad \int dx$$

If n is a positive fraction we can say that the problem is reduced to its simplest form.

We must now examine the special types on p. 86 more closely.

Detailed Study of Types of Integral

Type I. The principle here is to make a simple change in the constants which changes the numerator into the derivative of the denominator plus a constant. We can then integrate the first expression directly and are left with Type Ia.

We study
$$I = \int \frac{(Cx+D)\,dx}{ax^2+2bx+c}$$

Choose λ and μ so that
$$Cx+D \equiv 2\lambda(ax+b)+\mu$$

This gives
$$\lambda = \frac{C}{2a} \quad \text{and} \quad \mu = D-\frac{bC}{a}$$

Then an integral becomes
$$I = \frac{C}{2a}\int \frac{d(ax^2+2bx+c)}{ax^2+2bx+c} - \left(D-\frac{bC}{a}\right)\int \frac{dx}{ax^2+2bx+c}$$
$$= \frac{C}{2a}\log(ax^2+2bx+c) + \text{Type Ia}$$

In Type Ia we have
$$\int \frac{dx}{ax^2+2bx+c} = \frac{1}{a}\int \frac{dx}{\left(x^2+\frac{2b}{a}x+\frac{c}{a}\right)}$$
$$\frac{1}{a}\int \frac{dx}{\left(x+\frac{b}{a}\right)^2+\left(\frac{ca-b^2}{a^2}\right)}$$

and this is a Standard Form (8a) if $b^2 < ac$.

If $b^2 > ac$ the Standard Form is of the type 9.
$$\int \frac{dx}{x^2-a^2} = \frac{1}{2a}\log\frac{x-a}{x+a} \quad \text{if} \quad x^2 > a^2$$

or
$$\frac{1}{2a}\log\frac{(a-x)}{a+x} \quad \text{if} \quad x^2 < a^2$$

Type II can be similarly reduced to an integral which we can write down together with a remainder which is Type IIa. So we consider

$$\int \frac{dx}{(ax^2+2bx+c)^n}$$

It will clearly suffice if we can evaluate the corresponding integral

$$I_n = \int \frac{dt}{(t^2+\alpha^2)^n}$$

$$= \frac{1}{\alpha^2}\int \frac{dt}{(t^2+\alpha^2)^{n-1}} - \frac{1}{\alpha^2}\int \frac{t^2\,dt}{(t^2+\alpha^2)^n}$$

$$= \frac{1}{\alpha^2}\int \frac{dt}{(\alpha^2+t^2)^{n-1}} + \frac{1}{2\alpha^2(n-1)}\int t\frac{d}{dt}\left\{\frac{1}{(\alpha^2+t^2)^{n-1}}\right\} dt$$

Integrate by parts the second term and we have

$$I_n = \frac{1}{\alpha^2}\left[\frac{2n-3}{2(n-1)}\int \frac{dt}{(\alpha^2+t^2)^{n-1}}\right] + \frac{t}{2\alpha^2(n-1)(\alpha^2+t^2)^{n-1}}$$

$$= \frac{1}{\alpha^2}\left[\frac{2n-3}{2(n-1)}I_{n-1}\right] + \frac{t}{2\alpha^2(n-1)(\alpha^2+t^2)^{n-1}}$$

So we have found a reduction formula.

Type III

$$\int \frac{Cx+D}{\sqrt{(ax^2+2bx+c)}}$$

may clearly be reduced to

$$\frac{C}{2a}\int \frac{2(ax+b)\,dx}{\sqrt{(ax^2+2bx+c)}} + \frac{D-bC}{a}\int \frac{dx}{\sqrt{(ax^2+2bx+c)}}$$

$$= \frac{C}{a}\sqrt{(ax^2+2bx+c)} + \text{Type IIIa}$$

So we now consider

$$\int \frac{dx}{\sqrt{(ax^2+2bx+c)}}$$

Depending on the numerical values of *a*, *b*, *c*, this can be reduced to three forms

$$\int \frac{dx}{\sqrt{(\alpha^2-x^2)}}, \qquad \int \frac{dx}{\sqrt{(x^2+\alpha^2)}} \quad \text{or} \quad \int \frac{dx}{\sqrt{x^2-\alpha^2}}$$

The first is $\sin^{-1} x/\alpha$ and is Standard Form 7.

To integrate the second put $\sqrt{x^2+\alpha^2} = -x+t$

then $$\alpha^2 = -2xt+t^2$$

$$x = \frac{t^2-\alpha^2}{2t}, \qquad -x+t = \frac{(\alpha^2+t^2)}{2t}$$

and $$\frac{dx}{dt} = \frac{4t^2-(t^2-\alpha^2)2}{4t^2} = \frac{2t^2+2\alpha^2}{4t^2} = \frac{\alpha^2+t^2}{2t^2}$$

Whence $$\int \frac{dx}{\sqrt{(x^2+\alpha^2)}} = \int \frac{(\alpha^2+t^2)2t}{2t^2(\alpha^2+t^2)}\,dt = \int \frac{dt}{t}$$

$$= \log t = \log(x+\sqrt{x^2+\alpha^2})$$

Similarly $$\int \frac{dx}{\sqrt{(x^2-\alpha^2)}} = \log(x+\sqrt{x^2-\alpha^2})$$

These last two should be called Standard Forms 11 and 12.

If we now turn our attention to Type IIIb we have

$$I = \int \sqrt{ax^2+2bx+c}\,dx$$

By elementary algebra this can be reduced to one of the three equivalent forms

$$\int \sqrt{x^2+\alpha^2}\,dx, \qquad \int \sqrt{x^2-\alpha^2}\,dx, \qquad \int \sqrt{\alpha^2-x^2}\,dx$$

Taking the first and integrating by parts

$$\int \sqrt{x^2+\alpha^2}\,dx = x\sqrt{x^2+\alpha^2} - \int \frac{x^2\,dx}{\sqrt{x^2+\alpha^2}}$$

$$2\int \sqrt{x^2+\alpha^2}\,dx = x\sqrt{x^2+\alpha^2} + \alpha^2 \int \frac{dx}{\sqrt{x^2+\alpha^2}}$$

i.e. $$\int \sqrt{x^2+\alpha^2}\,dx = \frac{x}{2}\sqrt{x^2+\alpha^2} + \frac{\alpha^2}{2}\log\{x+\sqrt{x^2+\alpha^2}\}$$

Similarly we may obtain

$$\int \sqrt{x^2-\alpha^2}\,dx = \frac{x}{2}\sqrt{x^2-\alpha^2} - \frac{\alpha^2}{2}\log\{x+\sqrt{x^2-\alpha^2}\}$$

and $$\int \sqrt{\alpha^2-x^2}\,dx = \frac{x}{2}\sqrt{\alpha^2-x^2} + \frac{\alpha^2}{2}\sin^{-1}\frac{x}{\alpha}$$

Type IV

$$\int \frac{dx}{(x-\alpha)\sqrt{(ax^2+2bx+c)}}$$

put $$x-\alpha = \frac{1}{t} \quad \text{so} \quad \frac{dx}{x-\alpha} = -\frac{dt}{t}$$

and this reduces to the form

$$\int \frac{dt}{\sqrt{(a_1t^2+2b_1t+c_1)}} \quad \text{which is type IIIa.}$$

Let us now briefly consider types IVa and IVb. Using the substitution $x-\alpha = 1/t$ we can, as we have just shown, reduce IVb to IVa, so we need only consider IVa.

$$I_m = \int \frac{x^m\,dx}{\sqrt{ax^2+2bx+c}}$$

write $$y = ax^2+2bx+c$$

Now

$$\frac{d}{dx}(x^{m-1}\sqrt{y}) = (m-1)x^{m-2}\sqrt{y}+\frac{x^{m-1}(ax+b)}{\sqrt{y}} \tag{4.1}$$

$$= \frac{Ax^m+Bx^{m-1}+Cx^{m-2}}{\sqrt{y}}$$

where A, B, C are constants.

If we now integrate (4.1)

$$x^{m-1}\sqrt{y} = AI_m+BI_{m-1}+CI_{m-2}$$

This relates I_m to I_{m-1} and I_{m-2}

But the values of I_m for $m = 1$ and 0 are just Types III and IIIa, so we can perform the integration.

Finally we have Type V.

$$I = \int \frac{Ux+V}{(Ax^2+2Bx+C)\sqrt{(ax^2+2bx+c)}}$$

To simplify we try to reduce *both* quadratic factors to sums of squares. This is possible if we put

$$x = \frac{\mu t+v}{t+1}$$

and choose μ and v so that

$$a\mu v+b(\mu+v)+C = 0$$

and
$$A\mu v+B(\mu+v)+C = 0$$

If we carry out the necessary algebra we find that I can be reduced to the form

$$I = P\int \frac{t\,dt}{(\alpha t^2+\beta)\sqrt{\gamma t^2+\delta}}+Q\int \frac{dt}{(\alpha t^2+\beta)\sqrt{\gamma t^2+\delta}}$$

if we put $t = \frac{1}{u}$ the first of these reduces to the form of the second so we need only consider

$$I' = \int \frac{dt}{(\alpha t^2+\beta)\sqrt{\gamma t^2+\delta}}$$

we have to find a substitution which will rationalise this.

It is $\frac{t}{\sqrt{\gamma t^2+\delta}} = u$ and this leads to

$$I' = \int \frac{du}{\beta+(\alpha\delta-\beta\gamma)u^2}$$

and this is Type I.

The reader will have noticed, possibly with some slight irritation, that all we have done in general is to indicate that the types of integral in the last section can be integrated, and to point the way to reducing them to some Standard Form. We have not carried out the algebra required to obtain the correct answer.

This is in accordance with the philosophy expressed at the beginning of the chapter. We want the reader to be able after a little examination of the problem to be able to recognise the type and to see how to set about obtaining the answer. The rest becomes a matter of filling in detail, and we did say that the going could be tough.

Resumé of the Integrability of Algebraic Expressions

Polynomials

Since these are of the type $a_0x^n+a_1x^{n-1}+ \ldots + a_n$ they present no problem.

Rational Functions

Any rational function can be reduced to the form of a sum of terms of the type

$$\frac{A}{x-\alpha}, \quad \frac{B_r}{(x-\alpha)^r}, \quad \frac{Cx+D}{x^2+2bx+c}, \quad \frac{C_n x+D_n}{(x^2+2bx+c)^n}$$

where A, B_r, C, D, $C_n D_n$ are constants. We know how to deal with all these.

Algebraic Functions

Clearly the field here is very wide. We shall confine ourselves to the study of a rational function of x and $\sqrt{R}$ where $R = ax^2+2bx+c$, i.e. we study $F = F(x, \sqrt{R})$.

Since F must be the ratio of two polynomials we must have if we multiply out

$$F = \frac{A+B\sqrt{R}}{C+D\sqrt{R}} \qquad \text{A and B being polynomials}$$

We can then rationalise this to take the form

$$F = P+Q\sqrt{R}$$

P and Q are rational functions.

This leaves us effectively with $Q\sqrt{R}$, or the equivalent $Q'/\sqrt{R}$. This now can only give us integrals of the types

$$\int \frac{x^n\,dx}{\sqrt{R}}, \quad \int \frac{dx}{(x-\alpha)^m\sqrt{R}}$$

$$\int \frac{(p_r x+q_r)\,dx}{(\alpha x^2+2\beta x+\gamma)^r\sqrt{R}}$$

and we shall only consider the case $r = 1$.

All of these have been studied and the methods to obtain the answer have been given.

Integration of Transcendental Functions

As before we pass on to consider the integration of transcendental functions. These form such a large class that we can only consider a few of the more important cases where integrals can be found.

(1) Polynomials in sin x and cos x, e.g.

$$\int \sin^3 x \cos^2 x \, dx$$

The method is to express the power of sin x and cos x in terms of multiples of x.

In the one case we have

$$\sin^3 x = \tfrac{1}{4}(3 \sin x - \sin 3x)$$

$$\cos^2 x = \tfrac{1}{2}(1 - \cos 2x)$$

We now have

$$\tfrac{1}{8}\int [3 \sin x - \sin 3x - 3 \sin x \cos 2 + \sin 3x \cos 2x] \, dx$$

$$= \tfrac{1}{8}\int [3 \sin x - \sin 3x - \tfrac{3}{2}(\sin 3x - \sin x) + \tfrac{1}{2}(\sin 5x + \sin x)] \, dx$$

$$\tfrac{1}{16}\int \{10 \sin x - 5 \sin 3x + \sin 5x\} \, dx$$

$$-\tfrac{5}{8} \cos x + \frac{5 \cos 3x}{48} - \frac{\cos 5x}{80}$$

We could, of course, have done this another way:

$$\int \sin^3 x \cos^2 x \, dx = \int (1 - \cos^2 x)(\cos^2 x) \sin x \, dx$$

If we put $\cos x = t$ this becomes

$$-\int (1-t^2)t^2 \, dt = \frac{t^5}{5} - \frac{t^3}{3} = \frac{\cos^5 x}{5} - \frac{\cos^3 x}{3}$$

This can be reduced to the form above.

The integrals $\int x^n \cos x \, dx$ and $\int x^n \sin x \, dx$

can be found by a reduction formula

for $\int x^n \cos x \, dx = x^n \sin x \quad -n\int x^{n-1} \cos x \, dx$

so provided n is a positive integer we can find the answer.

It follows that we can also integrate any polynomial in x, $\cos x$, and $\sin x$, or more generally we can integrate

$$P(x, \cos ax, \sin bx).$$

Rational Function of sin *x* and cos *x*

If we make the substitution

$$\cos x = \frac{1-t^2}{1+t^2}, \qquad \sin x = \frac{2t}{1+t^2}\ \frac{dx}{dt} = \frac{2}{1+t^2}$$

then we reduce the problem to that of integrating a rational function of t.

Integrals Involving $\sin^{-1} x$, $\tan^{-1} x$, or $\log x$

Here we integrate by parts and obtain

$$\int \sin^{-1} x\, dx = x \sin^{-1} x - \int \frac{x\, dx}{\sqrt{1-x^2}}$$

$$= x \sin^{-1} x + \sqrt{1-x^2}$$

$$\int \tan^{-1} x\, dx = x \tan^{-1} x - \int \frac{x\, dx}{1+x^2}$$

$$= x \tan^{-1} x - \tfrac{1}{2} \log(1+x^2)$$

$$\int \log x\, dx = x \log x - \int dx = x(\log x - 1)$$

By means of reduction formulae we can evaluate polynomials in x and $\sin^{-1} x$, x and $\tan^{-1} x$, or x and $\log x$. E.g.

$$\int x^m (\log x)^n\, dx = \frac{x^{m+1}}{m+1} (\log x)^n - \frac{n}{m+1} \int x^m (\log x)^{n-1}\, dx$$

and by continuing we reduce this to $\int x^m\, dx$.

Examples

Integrate with respect to x.

1. $x \sin x$
2. $x\, e^{-x^2}$
3. $x \sin^{-1} x$
4. $\sqrt{2x+x^2}$
5. $\dfrac{x^2}{\sqrt{1-x^2}}$
6. $\dfrac{x}{(x^2+1)(x^2+3)}$
7. $\dfrac{\sqrt{x}}{1+x}$
8. $x^3\sqrt{x^2-2}$
9. $\dfrac{1}{x^4\sqrt{1+x^2}}$
10. $\sqrt{\tan x}$
11. If $u_n = \int x^n\sqrt{(1-x^2)}\, dx$

show that

$$u_n = \frac{-x^{n-1}(1-x^2)^{3/2}}{n+2} + \frac{n-1}{n+2}u_{n-2}$$

CHAPTER FIVE

INTEGRAL CALCULUS

(*Continued*)

Definite Integrals

Let $y = f(x)$ be a continuous function of x and let the graph be shown as in Figure 13.

Then the region bounded by the two ordinates AD, BE, and the portion of the x axis DE together with the portion

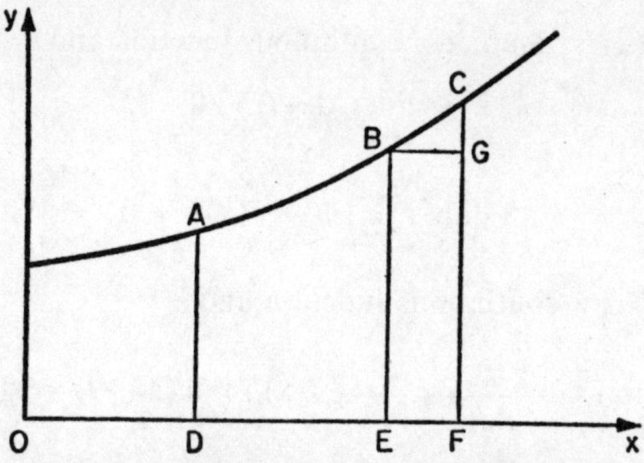

FIG. 13

of the curve $y = f(x)$ lying between A and B is called the *area* of the region and we assume that we can assign a measure to this. We cannot as yet give rules for finding this measure, but we do know how to find the measure of simple regions such as rectangles, triangles, etc.

If we suppose AD to be fixed but assume that B moves to C then the area alters and has now become the region $ACFD$. A little thought shows that the area is a function of x since the coordinates of the movable point B are

$(x, f(x))$ and of the point E (which moves with B) are (x, O).

Let the area be denoted by $F(x)$.

Let the coordinates of B be (x, y) and C be $(x+h, y+k)$, then

$$F(x+h)-F(x) = hf(x)+\text{area } BCG$$

Now, let M be the greatest value of $f(x)$ in the interval $(x, x+h)$. Let

$$FG = l$$

then the ordinate of any point on the curve BC measured from GB is always less than $M-l$. Hence the area $BCG < (M-l)h$.

But $f(x)$ is assumed a continuous function and in this case

$$\lim_{h\to 0} (M-l) = 0$$

Hence, (1)

$$\lim_{h\to 0}[F(x+h)-F(x)] = 0$$

i.e. $F(x)$ is a continuous function and

$$\lim_{h\to 0}\frac{F(x+h)-F(x)}{h} = f(x)+\lim_{h\to 0} (M-l) = f(x)$$

So

$$F'(x) = f(x)$$

This provides us with one rule for finding the measure of the area. We calculate the integral of $f(x)$. This will have an arbitrary constant. We may choose this so that $F(a) = 0$ if the point D is at $x = a$.

Then the measure of the area $ABDE$ is

$$F(x)-F(a) \tag{5.1}$$

Of course, as we are fully aware, it may not be possible to evaluate $F(x)$ explicitly so we have a solution using the integral sign to express (5.1).

We say

$$\text{Area } ABED = \int_a^x f(x)\,dx$$

We have so far thought of the area as a variable—a function of x the coordinate which fixes the moving point B. More usually we wish to calculate a definite area such as that lying between the ordinate at $x = a$ and $x = b$.

In this case clearly the area is

$$\text{Area} = \int_a^b f(x)\,dx$$

This, if we can find $F(x)$, is, by (5.1),

$$\text{Area} = F(b) - F(a)$$

The expression

$$\int_a^b f(x)\,dx$$

is called a *Definite Integral*, a and b are the *upper* and *lower* limits. Sometimes, to distinguish ideas, the function

$$\int f(x)\,dx = F(x)$$

is referred to as the indefinite integral of $f(x)$.

The following results, which are often useful, are direct and immediate consequences of the definition.

1. $\int_a^b f(x)\,dx = -\int_b^a f(x)\,dx$
2. $\int_a^b f(x)\,dx = \int_a^c f(x)\,dx + \int_c^b f(x)\,dx$
3. $\int_a^b hf(x)\,dx = h\int_a^b f(x)\,dx$
4. $\int_a^b \{f(x)+\varphi(x)\}\,dx = \int_a^b f(x)\,dx + \int_a^b \varphi(x)\,dx$

   If $f(x)$ lies between the values m and M in the interval (a, b) then

5. $m(b-a) \leqq \int_a^b f(x)\,dx \leqq M(b-a)$

   and hence

6. $\int_a^b f(x)\,dx = (b-a)f(\xi)$

When ξ is some number lying between (a and b) we can extend results 5 and 6 to the more general result:

7. If $\varphi(x)$ is positive in the interval (a, b) then

$$m \int_a^b \varphi(x)\, dx \leqq \int_a^b f(x)\varphi(x)\, dx \leqq M \int_a^b \varphi(x)\, dx$$

and $$\int_a^b \varphi(x)f(x)\, dx = f(\xi) \int_a^b \varphi(x)\, dx$$

The techniques of integration of parts and of substitution may be applied to the evaluation of definite integrals. Examples will illustrate these applications.

The reader should note that occasionally we can evaluate the definite integral when the indefinite integral cannot be found. This may seem a little strange, but it is, of course, basically simpler to find the *difference* $F(b)-F(a)$ between two particular values of $F(x)$ than it is to find $F(x)$ itself.

The Definite Integral as the Limit of a Sum

We have just seen how, faced with the problem of finding a number equal to the area of a figure which has a curve as part of its boundary, we have been able, thanks to our knowledge of the integral calculus, to give a rule for finding this number.

We noted that the rule was a formal one, and not necessarily one which can be practically applied in every case.

Note too that we used geometrical intuition in arriving at the rule, and that we did not prove that a number giving the area necessarily exists.

Proofs of existence theorems tend to be lengthy and some think them dull. It will be obvious, however, that cases arise in which intuition is no help and then existence theorems can be extremely important.

We wish to study the evaluation of definite integrals from the point of view of the limit of a sum. Not only does this give us greater insight into the meaning of an integral, but it is the logical foundation for the study of multiple integrals, for which geometrical ideas are not always applicable.

Let $f(x)$ be a function of x continuous in the interval (a, b) whose graph is as in Figure 14.

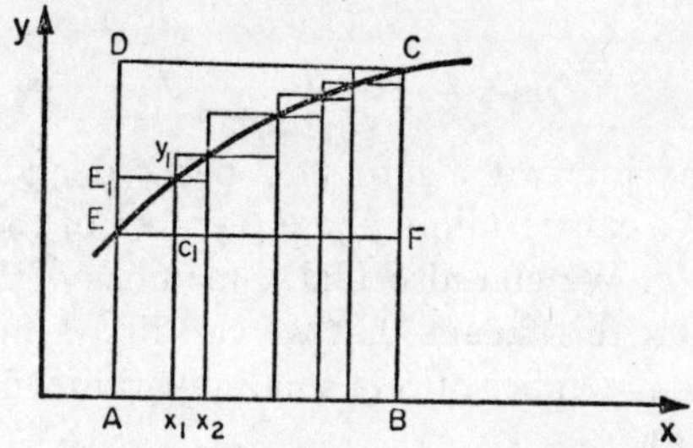

FIG. 14

Divide the base into a number of intervals, not necessarily equal, which we call $\delta x_1, \delta x_2 \ldots$ where δx_1 is the length of the interval (A, x_1) and δx_i the length of the interval (x_{i-1}, x_i), etc.

Erect perpendiculars as shown and complete the elementary rectangles such as $AEC_1\,x_1$ and $AE_1\,y_1x_1$.

In the figure $f(x)$ is shown as an increasing function but this is not essential.

Our object is to show that there is a number which corresponds to the area $AECB$, and that this number is the limit of the sums formed by adding up such rectangles as AEC_1x_1 or $AE_1y_1x_1$.

Proof. Let the sum of the rectangles such as AEC_1x_1 be s. By varying the method of subdivision we can form an aggregate of sums such as s. We shall show formally what is geometrically apparent, viz. that any number of the aggregate of sums s is less than the area $ABCD$. Hence the aggregate of sums s is bounded above and hence it has an upper bound j.

Similarly the aggregate of sums S formed by summing rectangles such as $AE_1y_1x_1$ in various modes, is bounded below and has a lower bound J.

It seems obvious geometrically that $j < J$, but let us see what this involves.

Suppose $\qquad j > J \quad \text{so} \quad j-\varepsilon = J$

Divide the interval ε into two parts μ and v so that $\mu+v = \varepsilon$. We can certainly find a member of the set s such that $s > j-\mu$. We can also find a member of S such that $S < J+v$. But this means that we can find a member of s greater than a member of S (a small diagram will make this appear obvious).

Hence to prove that $j < J$ we need only show that no member of the aggregate of s can be greater than any member of the aggregate of S.

This is easy to prove. For take two methods of division which give aggregates s and S, on the one hand, and s' and S' on the other. Combine these methods to give the sums $\bar{s}$ and $\bar{S}$. Now it is obvious that the combined method has more subdivisions than either that forming s or s'.

Suppose that a rectangle used in calculating s becomes two rectangles in calculating $\bar{s}$. Because of the method of construction the sum of the two smaller rectangles is certainly not less than the area of the single one.

From this it follows that $\qquad \bar{s} \geqq s$

Similarly $\qquad \bar{s} \geqq s'$

By precisely the same arguments we may show that

$$\bar{S} \leqq S$$

and $\qquad \bar{S} \leqq S'$

but $\qquad \bar{s} < \bar{S}$

and so we have the relationship

$$s' \leqq \bar{s} < \bar{S} \leqq S$$

Whence, since the method of forming s and S was arbitrary, we have shown that any sum such as s is less than any sum S'. This shows that $j < J$.

If we can show that, in the limit when the length of the subintervals are made sufficiently small, $j = J$ then we have proved our result.

We shall now set down the proof formally.

Let $f(x)$ be continuous in the interval (a, b) and let M and m be the upper and lower bounds of $f(x)$ in the interval. Divide the interval into sub-intervals δ_i and let M_i and m_i be the upper and lower bounds of $f(x)$ in the sub-interval δ_i. Let ξ_i be any point in the interval. Then the relationship below must hold

$$m\delta_i \leqq m_i\delta_i \leqq f(\xi_i)\delta_i \leqq M_i\delta_i \leqq M\delta_i$$

Summing over all the intervals

$$m(b-a) \leqq \sum m_i\delta_i \leqq \sum f(\xi_i)\delta_i \leqq \sum M_i\delta_i \leqq M(b-a)$$

Write $\qquad s = \sum_i m_i\delta_i \quad \text{and} \quad S = \sum M_i\delta_i$

Both s and S depend on the method of subdivision chosen, and we can form an aggregate of s and S by choosing various methods. The aggregate of all possible s's is bounded since all s's are less than $M(b-a)$.

Similarly the S's form a bounded aggregate for all S's are greater than $m(b-a)$.

Hence the aggregate of the s's has an upper bound j, and the aggregate of the S's has a lower bound J.

We must next consider the value of $S-s$.

This is

$$S-s = \sum (M_i - m_i)\delta_i$$

It is thus equal to the oscillation of the function $f(x)$ in the interval δ_i times δ_i and summed over all such intervals.

We must now consider the value of the expression $S-s$ in the limiting case when all intervals are less than some definite size δ.

First, consider the itentity

$$S-s = (S-J)+(J-j)+(j-s)$$

All the quantities in brackets are positive, so that if we can make $S-s$ less than some given quantity ε then each of them is $< \varepsilon$ and since $J+j$ are constants the only conclusion is that $J-j = 0$. This would prove our result.

The question is this. Can we by choosing all intervals $< \delta$ make $S-s < \varepsilon$.

This can only be proved by using a theorem—the Heine-Borel Theorem (or some equivalent) about intervals on a line.

We shall indicate the point at issue. For a proof the reader is referred to Hardy, *Pure Mathematics*, p. 196.

The Heine-Borel Theorem

If we have an interval (a, b) and we can define a set of intervals I, each of whose members lies in (a, b) with the following properties:

(1) Every point ξ of the interval is *inside* at least one interval of the set I.

(2) a is the end point of at least one interval of the set and b is the end point of at least one interval of the set.

Then we can cover the interval (a, b) with a *finite* number of intervals from the set.

The important point of the theorem is the word *finite*.

Now, in the case of a continuous function $\varphi(x)$, if we take any point ξ in (a, b) we can choose an interval $(\xi-\eta, \xi+\eta)$ which has the property that in this interval the oscillation of

φ is less than a given number ε. η depends on ξ but any quantity $< \eta$ will also do. Hence for each ξ we have a set of intervals with the required property that the oscillation is $< \varepsilon$. Now take all such intervals associated with all ξ of the interval (a, b).

These form a set with the properties required for the Heine-Borel Theorem. Hence there is a *finite* number of intervals which cover (a, b) in each of which the oscillation is $< \varepsilon$. These intervals are not necessarily equal in length.

Now so choose a set of intervals, but this time take $\varepsilon/2$ as the limit of oscillation.

From the intervals take the smallest δ.

Divide the (a, b) into intervals $< \delta$.

Then since each of these new intervals either be entirely inside the original set or, at most, inside two neighbours, the oscillation cannot exceed $2 \times \varepsilon/2$ and therefore $< \varepsilon$.

We wish to show that *given* any arbitrary small number ε we can choose a number δ so that, if all the intervals δ_i in our arrangement are *less than* δ then

$$S - s < \varepsilon$$

Now ε is *given*. From the number $\varepsilon/b-a$ we know that we can find a number δ such that if an interval (a, b) is divided into sub-intervals each $< \delta$ then the oscillation in *each of these* is $\varepsilon/b-a$,

i.e. $M_i - m_i < \varepsilon/b-a$ for all δ_i provided all $\delta_i < \delta$

So choose the δ_i then

$$S - s = \sum (M_i - m_i) < \frac{\varepsilon}{b-a} \sum \delta_i < \varepsilon$$

Multiple Integrals

The concept of an integral as the limit of a sum is easy to extend to cover the case of multiple integrals.

We shall study the case of double integration; the extension to three or more variables is easy to understand.

Just as the concept of an area proved convenient when studying simple integrals, so the idea of a volume is helpful in the case of double integration. Consider Figure 15.

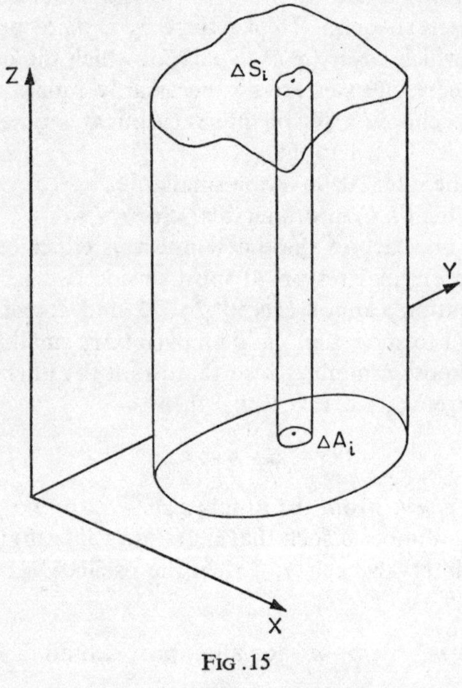

FIG. 15

Let $$z = f(x, y) \quad \text{be a surface}$$

Let ΔA_i be a small element of area in the XY plane and on it erect a cylinder perpendicular to this plane so as to cut the surface $z = f(x, y)$ in a surface element ΔS_i.

Let (ξ_i, η_i) be a point inside ΔA_i and let ζ_i be the corresponding point on S. Of course $\zeta_i = f(\xi_i, \eta_i)$.

Let M_i, m_i be the maximum and minimum values of ζ_i within the area ΔA_i.
Define the sums

$$S = \sum M_i \Delta A_i$$
$$s = \sum m_i \Delta A_i$$
$$V = \sum \zeta_i \Delta A_i$$

Then clearly $$s \leqq V \leqq S$$

We can now proceed to prove the existence of V by showing that S and s have the same limit when the areas A tend to zero. Hence V also has a limit which is in fact the volume of the cylinder enclosed by the surface $z = f(x, y)$ and the closed area on the XY plane.

A formal proof follows the lines of that given previously for a simple integral.

We write the expression for V as

$$V = \int f(x, y)\, dA$$

When we proceed to the actual evaluation of the integral, let us first take the case where $\Delta A = \Delta x_i \Delta y_i$. This means

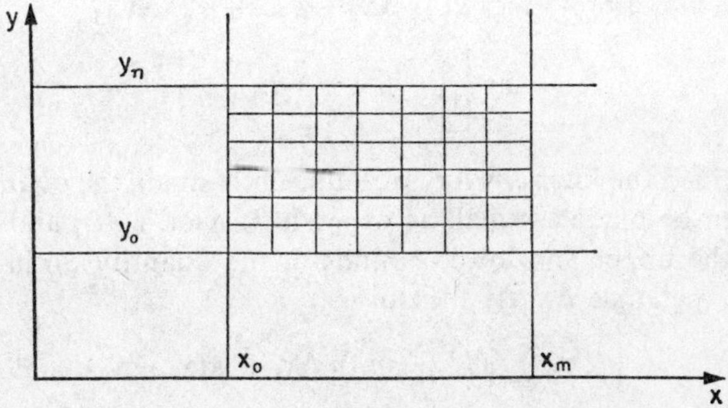

FIG. 16

that we divide the area on the XY plane into elementary rectangles $\Delta x_i \Delta y_i$ as in Figure 16. For further simplicity we suppose that the area on the XY plane is also rectangular and bounded by the lines

$$x = x_0, \qquad x = x_m \quad \text{and} \quad y = y_0, \quad y = y_n$$

We now wish to evaluate the sum

$$\sum \zeta_{ij} \Delta x_i \Delta y_j$$

where ζ_{ij} is the value of $f(x, y)$ at a point in the interior of the rectangle $\Delta x_i \Delta y_j$. Let there be m columns and n rows so that

$$\Delta x_i = x_i - x_{i-1} \quad \text{and} \quad \Delta y_y = y_y - y_{y-1}$$

We may then write the summation in full as

$$\begin{aligned}
&\zeta_{11} \Delta x_1 \Delta y_1 + \zeta_{21} \Delta x_2 \Delta y_1 + \zeta_{31} \Delta x_3 \Delta y_1 + \ldots \\
&\qquad\qquad + \zeta_{m1} \Delta x_m \Delta y_1 \\
&+ \zeta_{12} \Delta x_1 \Delta y_2 + \zeta_{22} \Delta x_2 \Delta y_2 + \zeta_{32} \Delta x_3 \Delta y_2 + \ldots \\
&\qquad\qquad + \zeta_{m2} \Delta x_m \Delta y_2 \\
&+ \zeta_{1n} \Delta x_1 \Delta y_n + \zeta_{2n} \Delta x_2 \Delta y_n + \zeta_{3n} \Delta x_2 \Delta y_n + \ldots \\
&\qquad\qquad + \zeta_{mn} \Delta x_m \Delta y_n
\end{aligned}$$

The sum of row j is

$$\begin{aligned}
&\Delta y_j [\zeta_{1j} \Delta x_1 + \zeta_{2j} \Delta x_2 + \ldots + \zeta_{mj} \Delta x_m] \\
&\qquad = \Delta y_j \left[\int_{x_0}^{x_m} f(x, y_j)\, dx + \varepsilon_j \right]
\end{aligned}$$

Provided the intervals Δx_i are sufficiently small, the quantity ε_j can be made as small as we wish. In fact if M_{ij} and m_{ii} are the upper and lower bounds of the quantity ζ_{ij} in the little rectangle $\Delta x_i \Delta y_j$, then

$$|\varepsilon_j| \leqq \sum_i |(M_{ij} - m_{ij})| \Delta x_i \leqq \kappa(x_m - x_0)$$

where κ is the maximum value of the variation $(M_{ij} - m_{ij})$ over any of the rectangles.

Let us now sum up the rows, and we get

$$\Delta y_1 \int_{x_0}^{x_m} f(x, y_1)\, dx + \Delta y_2 \int_{x_0}^{x_m} f(x, y_2)\, dx + \ldots + \Delta y_n$$

$$\times \int_{x_0}^{x_m} f(xy_n)\, dx + \varepsilon_1 \Delta y_1 + \varepsilon_2 \Delta y_2 + \ldots + \varepsilon_n \Delta y_n$$

Let $$\int f(x, y_1)\, dx = \phi(y_1) \quad \text{etc.}$$

and $$|\varepsilon_1 \Delta y_1 + \varepsilon_2 \Delta y_2 + \varepsilon_3 \Delta y_3 + \ldots|$$
$$\leqq \kappa(x_m - x_0)(y_n - y_0) \leqq \lambda$$

So our sum becomes

$$\phi(y_1) \Delta y_1 + \phi(y_2) \Delta y_2 + \ldots + \phi(y_n) \Delta y_n + \lambda$$
$$= \int_{y_0}^{y_n} \phi(y)\, dy + \mu + \lambda$$

where μ, λ can also be arbitrarily small if the intervals Δy and Δx are small.

Hence our double sum becomes in the limit

$$\int_{y_0}^{y_n} \int_{x_0}^{x_m} f(x, y)\, dx\, dy \tag{5.2}$$

Had we reversed our order of summation we should have had the result

$$\int_{x_0}^{x_m} \int_{y_0}^{y_n} f(x, y)\, dy\, dx \tag{5.3}$$

The procedure for evaluating a double integral such as (5.3) is to first integrate with regard to y keeping x as a constant, and then to integrate the result with regard to x. Provided certain conditions are satisfied the order of integration is immaterial.

Physical Interpretation

It is helpful to associate multiple integrals with physical concepts. We have already considered the interpretation of $\iint g(x, y)\, dx\, dy$ as a volume.

This integral could also be interpreted as giving us the mass of a plane sheet of variable density ρ given by

$$M = \iint \rho(x, y)\, dx\, dy$$

In the same way the mass of a solid of variable density ρ which is now a function of x, y, z, can be expressed as

$$M = \iiint \rho(x, y, z)\, dx\, dy\, dz$$

To take a further example, the Moment of Inertia of an element of mass m about a given axis is defined as the product mr^2 where r is the distance from the axis.

When we wish to calculate the moment of inertia of a solid body, such as a sphere, for example, we have to use multiple integration. Thus the moment of inertia of a sphere of constant density ρ about a diameter can be expressed

$$\iiint \rho(x^2+y^2+z^2)\, dx\, dy\, dz$$

We obtain this integral simply by writing down the expression for the moment of inertia of an element of the sphere centred at the point (x, y, z), and this by definition is

$$\rho\, dx\, dy\, dz\, (x^2+y^2+z^2)$$

Now we sum over all such elements by means of multiple integration getting the result above. Naturally ρ could be variable which would complicate the integration but leave the actual expression to be operated on formally the same.

Change of Variable in Multiple Integration

The evaluation of a multiple integral can frequently be greatly simplified by a change of variable. This is especially true if the boundary of the region over which we calculate the integral is more simply expressed in terms of the new coordinates.

In two dimensions we frequently use circular polar coordinates. In this system the rectangular element of area

$dx\,dy$ is replaced by the element $r\,dr\,d\theta$ formed by the curvilinear elements in Figure 17.

Area shaded is approximately $r\,d\theta \times dr$

In this case $\iint f(x, y)\,dx\,dy$ becomes $\iint f(r, \theta)r\,dr\,d\theta$.

In three dimensions we may have cylindrical co-ordinates r, θ, z, or spherical polar coordinates r, θ, ϕ. In cylindrical

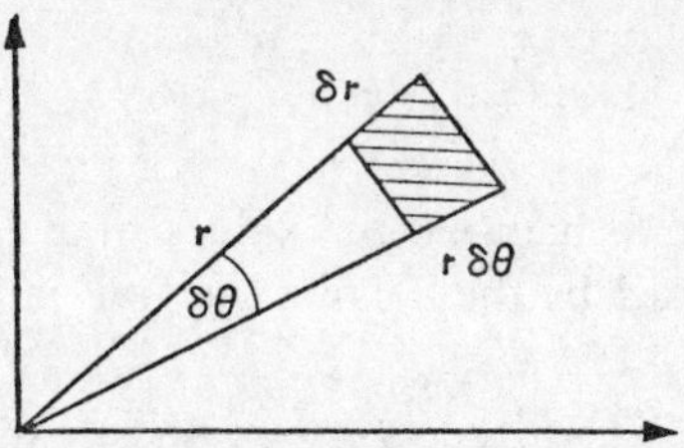

FIG. 17

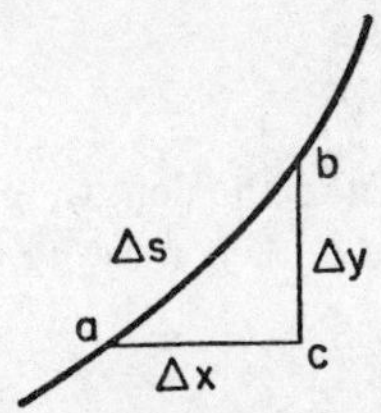

FIG. 18

coordinates the volume element is $r\,d\theta\,dr\,dz$, and in spherical polars it is $r^2 \sin\phi\,dr\,d\theta\,d\phi$, and we have to evaluate integrals.

$$\iiint f(r, \theta, \phi)r^2 \sin\phi\,dr\,d\theta\,d\phi \quad \text{or} \quad \iiint f(r, \theta, z)r\,dr\,d\theta\,dz$$

The reader will find practice in the examples.

The Length of a Plane Curve

Although it may not appear so, the idea of length in the case of a curve is a very difficult one. We have proved the existence of a *number* corresponding to the area of a plane figure. It is possible to prove that there is a number corresponding to the length of a curve but it is beyond the scope of this book. We can see easily enough what the expression for an element of length should be.

Consider Figure 18.

If we take an element of length Δs of a curved line whose graph is

$$y = f(x)$$

approximately $\quad (\Delta s)^2 = (\Delta x)^2 + (\Delta y)^2$

if we consider the area *abc* as approximately a right-angled triangle.

In the limit, using the terminology of differentials

$$ds = \sqrt{(\mathrm{dx})^2 + (dy)^2}$$

$$= \sqrt{1 + \left(\frac{dy}{dx}\right)^2}\, dx$$

Hence the length of a curve between the values of $x = a$ and $x = b$ may be expressed by the definite integral

$$l = \int_a^b \sqrt{1 + \left(\frac{dy}{dx}\right)^2}\, dx \qquad (5.4)$$

Example. For a circle of radius r

$$x^2 + y^2 = r^2 \quad \text{and} \quad 1 + \left(\frac{dy}{dx}\right)^2 = \frac{r^2}{r^2 - x^2}$$

The arc length from $x = 0$ to $x = r$ is

$$r\int_0^r \frac{dx}{\sqrt{r^2 - x^2}} = r\left[\sin^{-1}\left(\frac{x}{r}\right)\right]_0^r = \frac{\pi r}{2}$$

whence the circumference of a circle is $2\pi r$.

Some Geometrical Notions

Let *OP* be a line in space of three dimensions represented by Figure 19.

Let the angles measured between *OP* and the three axes *OX*, *OY*, *OZ* be θ, ϕ, ψ respectively.

Let
$$l = \cos\theta$$
$$m = \cos\phi$$
$$n = \cos\psi$$

then l, m, n are called the Direction Cosines of the line *OP*. l, m, n are not independent. In fact

$$l^2+m^2+n^2 = 1$$

To show this, consider a line of unit length. Its projections on the three coordinate axes are $1.\cos\theta$, $1.\cos\phi$, $1.\cos\psi$

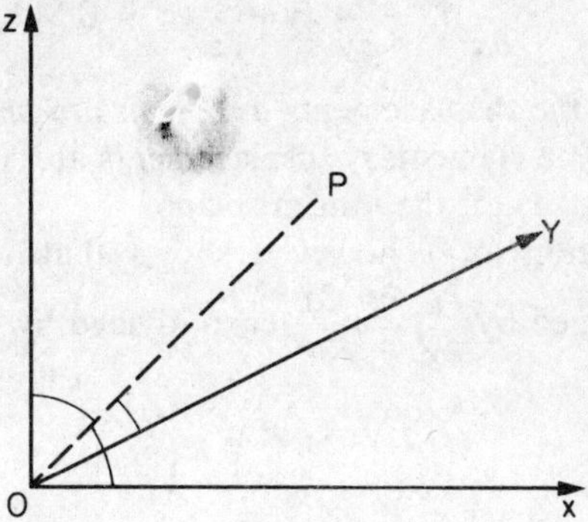

FIG. 19

or l, m, n. By the theorem of Pythagoras extended for three dimensions we can show that

$$l^2+m^2+n^2 = 1 \tag{5.5}$$

We can also show that if we have two lines with direction cosines l_1, m_1, n_1, and l_2, m_2, n_2 then the angle between them, θ, is given by

$$\cos\theta = l_1l_2+m_1m_2+n_1n_2 \tag{5.6}$$

Hence if the lines are perpendicular

$$l_1l_2+m_1m_2+n_1n_2 = 0$$

If a, b, c are numbers *proportional* to direction cosines we can find the line direction cosines by dividing a, b, c by $\sqrt{a^2+b^2+c^2}$ for then (5.5) is satisfied.

We are now in a position to define the direction of the

tangent plane at a point on a surface. All we need to do is to specify the direction cosines of the normal to the tangent plane.

Suppose we have a surface defined by

$$f(x, y, z) = \text{const.}$$

then

$$\frac{\partial f}{\partial x}\,dx+\frac{\partial f}{\partial y}\,dy+\frac{\partial f}{\partial z}\,dz = 0 \tag{5.7}$$

where here the displacements *dx*, *dy*, *dz* are *on* the surface, i.e. these three elementary displacements are not independent but must be on the tangent plane.

The relation (5.7), however, shows that the direction cosines defined by $\frac{\partial f}{\partial x}, \frac{\partial f}{\partial y}, \frac{\partial f}{\partial z}$ (each divided by

$$\sqrt{\left(\frac{\partial f}{\partial x}\right)^2+\left(\frac{\partial f}{\partial y}\right)^2+\left(\frac{\partial f}{\partial z}\right)^2}$$

if necessary) must therefore give the direction *normal* to the tangent plane since they define a direction normal to that given by *dx*, *dy*, *dz*.

Surface Areas

If now we take an element of surface δS and project this on to the XY plane, the normal to this element of surface makes an angle y with the Z azis where $\cos y = n$. Hence if the projection on the XY plane is δA we have

$$\delta A = \delta S \cos y$$

or

$$\delta S = \delta A \sec y$$

$$= \delta A \times \frac{\sqrt{\left(\frac{\partial f}{\partial x}\right)^2+\left(\frac{\partial f}{\partial y}\right)^2+\left(\frac{\partial f}{\partial z}\right)^2}}{\frac{\partial f}{\partial z}}$$

If we define our surface as

$$z = f(x, y)$$

then f_z' becomes 1 and we have

$$dS = \sqrt{1+\left(\frac{\partial z}{\partial x}\right)^2+\left(\frac{\partial z}{\partial y}\right)^2}\,dx\,dy$$

hence
$$S = \iint \sqrt{1+\left(\frac{\partial z}{\partial x}\right)^2+\left(\frac{\partial z}{\partial y}\right)^2}\,dx\,dy \qquad (5.8)$$

This expression, integrated between appropriate limits, enables us to calculate areas on a curved surface defined by the relation

$$z = f(x, y)$$

Examples

1. $\displaystyle\int_a^b x^4\,dx$

2. $\displaystyle\int_0^{\pi/2} \frac{\sin x\,dx}{2-\cos^2 x}$

3. $\displaystyle\int_0^{\pi} \frac{dx}{1-6\cos x+9}$

4. $\iint (x^2+y^2)\,dx\,dy$ taken over the sector of a circle of radius a bounded by the axes of x and y.

   [Hint: this may be written in the form

   $$\int_0^a dx \int_0^{\sqrt{a^2-x^2}} (x^2+y^2)\,dy\Big]$$

5. Show that the moment of inertia of a uniform sphere of mass M and radius a about an axis through the centre is $\frac{2}{5}Ma^2$.

6. Find the area formed by the portion of the surface given by

$$z = x^2 + y^2$$

cut off by the right cylinder given by

$$x^2 + y^2 = a^2$$

CHAPTER SIX

INFINITE SERIES AND INTEGRALS

Infinite Series

Let $u(n)$ be a function of n. If we add up a series of functions $u(n)$ for a finite series of values of n, say 1, 2, . . . N, we obtain a function for this sum

$$S(N) = u(1)+u(2)+u(3)+ \ \ldots \ u(N)$$

To save writing this is often expressed

$$S_N = u_1+u_2+u_3+ \ \ldots \ + u_N$$

or
$$S_N = \sum_{v=1}^{N} u_v$$

the Greek capital letter $\sum_{1}^{N}$ standing for the operation of carrying out the sum of N terms.

Now suppose we consider the case where the number of terms in the sum increase indefinitely, i.e. suppose

$$N \rightarrow \infty$$

Then if the function S_N has a limit S when $N \rightarrow \infty$ we can write

$$\lim_{n \to \infty} S_N = S$$

or
$$\lim_{n \to \infty} \sum_{v=1}^{n} u_v = S$$

In these circumstances we say that the series $\sum_{v=1}^{\infty} u_v$ *converges* to the sum S.

We may think of this as meaning that as we add more and more terms to the series we get closer and closer to the sum S.

If, on the other hand, the sum S_N tends to $+\infty$ or to $-\infty$ when N tends to infinity, then we say that the series *diverges*.

If the function S_N does not tend to a limit or to infinity when $N \to \infty$ then we say that the series oscillates finitely or infinitely.

The reader should note that, so far, in introducing the idea of infinite series we have used *no new ideas* other than those developed in the earlier chapters. They should be studied again very carefully if this is not obvious.

Some General Theorems on Infinite Series

I If the series $u_1+u_2+u_3+ \ldots$ converge to the sum S, then the expression $a+u_1+u_2+u_3+ \ldots$ converges to the sum $a+S$. This follows from the definition.

II If the series $u_1+u_2+u_3+ \ldots$ converges to the sum S, then the series $ku_1+ku_2+ku_3+ \ldots$ converges to the sum kS. This also follows from the definition.

III If the series $u_1+u_2+u_3+ \ldots$ converges to the sum S and the series $v_1+v_2+v_3+ \ldots$ converges to the sum V, then the series $(u_1+v_1)+(u_2+v_2)+(u_3+v_3) + \ldots$ converges to the sum $V+S$. This too follows from the definition.

The following theorems are less obvious and require proof.

IV If $u_1+u_2+u_3 + \ldots$ converges then $\lim\limits_{N-\infty} u_N = 0$

Proof. $$u_N = S_N - S_{N-1}$$

$$\therefore \quad \lim u_N = \lim (S_N) - \lim (S_{N-1}) = S - S = 0.$$

It is extremely important to note that the converse is NOT true. E.g. the series

$$1+\tfrac{1}{2}+\tfrac{1}{3}+ \ldots \quad +1/n+ \ldots$$

is *divergent* although the nth term $= 1/n$ tends to zero.

V If the series $u_1+u_2+u_3+\ \ldots$ is convergent, then a series formed out of groups of this series in brackets is also convergent and has the same sum.

E.g. if $u_1+u_2+u_3+u_4+\ \ldots$ is convergent with sum S, then the series

$$(u_1+u_2)+(u_3+u_4)+(u_5+u_6)+\ \ldots$$

is also convergent and has the sum S.

The proof is readily deducible from the definition. Here again the converse is NOT true for the series $1-1+1-1+\ \ldots$ oscillates whereas the series $(1-1)+(1-1)+(1-1)\ +\ \ldots$ is convergent with $S = 0$.

In what follows we assume that all the terms in the series are positive. This makes the theory simpler and the discussion can be extended later to cover cases where series contain negative terms.

VI If all the terms of a series are positive then the series must converge or diverge to $+\infty$. This follows at once from the fact that

$$S_n = u_1+u_2+\ \ldots\ +u_n$$

is an increasing function of n.

VII This theorem is of the utmost importance.

Let $\quad u_1+u_2+u_3+\ \ldots \quad$ and $\quad v_1+v_2+v_3+\ \ldots$

be two series of positive terms, and let $u_n \leqq kv_n$ for all values of n where k is a constant, then if the series $v_1+v_2+v_3\ \ldots$ is convergent, so is the series

$$u_1+u_2+u_3+\ \ldots$$

and its sum is less than or equal to the sum of

$$k(v_1+v_2+v_3+\ldots)$$

The proof is straightforward.
By definition

$$v_1+v_2+v_3+\ldots = t$$

then $v_1+v_2+v_3+\ldots+v_n < t$ for all values of n. But this means that $u_1+u_2+u_3+\ldots+u_n < kt$ for all values of n and this proves the theorem.
It is readily deducible that if $\sum U_n$ is divergent and if $V_n \geqq kUn$, then $\sum V_n$ is also divergent.

The Geometric Series

Clearly if the above theorem, which enables us to compare one series with another, is to be of much use, we need some standard series as a basis for such comparisons. A useful one is the 'geometric series' for which $u_n = r^{n-1}$ and

$$S_n = 1+r+r^2+r^3+\ldots+r^{n-1}$$

By elementary algebra

$$S_n = \frac{1-r^n}{1-r}$$

except where $r = 1$ when $S_n = n$. In this last case, of course,

$$S_n \to \infty$$

In the general case, S_n can only tend to a limit if r^n does. We have already shown that the condition for this is

$$-1 < r < 1$$

Thus we may assert that the geometric series

$$S = 1+r+r^2+r^3+\ldots$$

converges to the limit $1/(1-r)$ provided $-1 < r < 1$.

If $r \geqq 1$ it diverges, if $r = -1$ it oscillates finitely, and if $r < -1$ it oscillates infinitely.

The Ratio Test of Convergence

Let $u_1, u_2, u_3 \ldots u_n \ldots$ be a series of positive terms.

Then if
$$\lim_{n\to\infty} \frac{u_{n+1}}{u_n} = \rho$$

The series converges if $\rho < 1$.
The series diverges if $\rho > 1$.
If $\rho = 1$ the test tells us nothing.

Proof. If $\rho < 1$ then there must be some number r between ρ and 1 such that for all values of $n \geqq N$

$$\frac{u_{n+1}}{u_n} < r$$

then $\quad u_{n+1} < ru_n, \quad u_{n+2} < ru_{n+1} < r^2u_n, \quad$ and so on.

Then
$$u_n + u_{n+1} + u_{n+2} \ldots < u_n(1+r+r^2 \ldots)$$
$$< u_n \frac{1}{1-r}$$

since $r < 1$.

Hence the series converges.

The proof that the series diverges for $\rho > 1$ is left as an exercise.

Van Raabe's Test for Convergence

If the ratio test above fails there is a more delicate test which may be applied. This test is due to Van Raabe. The proof is beyond the scope of this book but will be found in the references.

If $\lim_{n\to\infty} \left[n \left(\frac{u_{n+1}}{u_n} - 1 \right) \right] > 1 \qquad$ the series converges

If $\lim_{n\to\infty}\left[n\left(\frac{u_{n+1}}{u_n}-1\right)\right] < 1$ the series diverges

If the limit is equal to 1 the test gives no indication.

The most useful test and the easiest to apply is the comparison test with the geometric series, i.e. the Ratio Test above. The fact is, however, that geometric series converge very rapidly and the comparison is not very useful for the large class of series which converge slowly.

Abel's Theorem

There is a useful theorem due to Abel which gives a test for divergence.

If $\sum u_n$ is a convergent series of positive and decreasing terms, then

$$\lim_{n\to\infty} (nu_n)=0$$

Suppose $\lim_{n\to\infty} (nu_n)$ is not zero.

Proof.

Then there are an infinite number of values of n for which

$$nu_n \geqq \sigma \tag{6.1}$$

Let $n_1, n_2, n_3, \ldots$ be a sequence of values of n for which (6.1) holds and which are also such that

$$n_2 > 2n_1, \qquad n_3 > 2n_2, \qquad \text{etc.}$$

Now, since u_n is a decreasing function of n

$$u_0+u_1+\ldots+u_{n_1-1} \geqq n_1u_{n_1} \geqq \sigma$$

$$u_{n_1}+\ldots+u_{n_2-1} \geqq (n_2-n_1)u_{n_2} \geqq \tfrac{1}{2}n_2u_{n_2} \geqq \sigma/2$$

$$u_{n_2}+\ldots+u_{n_3-1} \geqq (n_3-n_2)u_{n_3} \geqq \tfrac{1}{2}n_3u_{n_3} \geqq \sigma/2$$

Hence we can split up the series into groups of terms each

of which is $\geqq \sigma/2$. Hence the sum tends to infinity and the series is divergent.

As a useful example take

$$S = \sum_{1}^{\infty} \frac{1}{n}$$

Hence $\qquad u_n = 1/n \quad \text{and} \quad \lim nu_n = 1$

So the series is divergent.

The next general theorem of importance is called

Dirichlet's Theorem

The sum of a series of positive terms is the same in whatever order we arrange the terms.

For let the series be

$$u_1+u_2+u_3+ \ldots + u_n+ \ldots$$

and suppose we rearrange the terms so as to form the series

$$v_1+v_2+v_3+ \ldots + v_n+ \ldots$$

Since $\sum u_n$ is convergent, let the sum be S. Then the sum of any number of u's is $< S$. But every u is a v and so the sum of any number of v's is $< S$. Hence the series of v's is convergent. Let its sum be V. Then by the above

$$V \leqq S$$

But if we now start with the convergent series of v's we can show

$$S \leqq V$$

Hence $\qquad V = S.$

An important consequence of Dirichlet's Theorem is that we can give a rule for multiplication of two convergent series. Let $u_0+u_1+u_2+ \ldots$ and $v_0+v_1+v_2+ \ldots$ be two convergent series whose sums are U and V respectively. Then the series

$$u_0v_0+(u_1v_0+u_0v_1)+(u_2v_0+u_1v_1+u_0v_2)+\ \dots \qquad (6.2)$$

is convergent to the sum UV.

To prove this, form an array of the possible products.

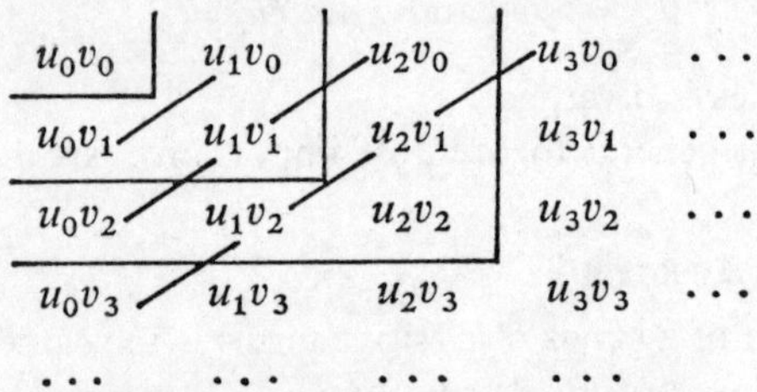

Now if we take groups by diagonal lines as shown we have the series we have written down.

But if we rearrange in another way, taking the terms in the rectangular areas, we obtain

$$u_0v_0+(u_0v_1+u_1v_0+u_1v_1)$$
$$+(u_0v_2+u_1v_2+u_2v_2+u_2v_1+u_2v_0)+$$

but this series is just

$$u_0v_0+[(u_0+u_1)(v_0+v_1)-u_0v_0]$$
$$+(u_0+u_1+u_2)(v_0+v_1+v_2)-(v_0+u_1)(v_0+v_1)$$

and the sum of $n+1$ groups is

$$(u_0+u_1+\ \dots\ +\ u_n)(v_0+v_1+\ \dots\ +\ v_n)$$

This limit of this sum is UV.

But as the second way of arranging the series is merely a rearrangement of the first, the first series (6.2) has the sum UV.

As we mentioned previously, the Comparison Test for convergence has so far been based on the geometric series which are types of series which converge very rapidly. They are therefore not sufficiently delicate for several types of convergent series.

If we make a further assumption—that the terms of the series are not only positive but also *decreasing*—then we can prove a theorem due to MacLaurin which uses an integral. We should mention that there is, in fact, no loss of generalisation supposing the terms of the series are decreasing, since by Dirichlet's Theorem we are at liberty to rearrange them to produce this property. We shall assume that this, if necessary, has been done.

Cauchy's Integral Test of Convergence

Let $u_n = f(n)$ and let $f(x)$ be a continuous steadily decreasing function of x which assumes the value $f(n)$ when n is an integer. Let us study the graph of the function $f(x)$

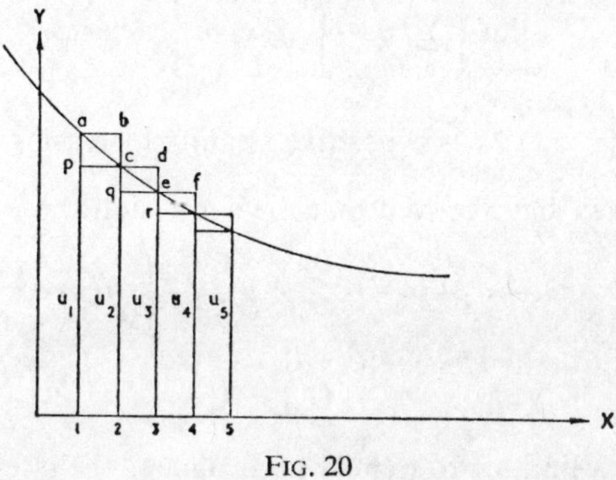

FIG. 20

as in Figure 20. We can interpret the terms $u_1, u_2, u_3 \ldots$ as the areas of rectangles such as $12ab$.

E.g. $$u_1 = \text{area } 12ab$$

Now define $$v_r = f(r-1) - \int_{r-1}^{r} f(x)\,dx$$

so $$v_3 = u_2 - \int_2^3 f(x)\,dx = \text{area cde}$$

Now $$v_r = \int_{r-1}^{r} \{f(r-1)-f(x)\}\,dx$$

$$0 \leqq v_r \leqq f(r-1)-f(r)$$

e.g. $$0 \leqq v_3 \leqq \text{area cdeq}$$

Hence $v_2+v_3+v_4+v_5+ \dots +v_n \leqq f(1)-f(n) \leqq f(1)$

So $$\sum_1^{n-1} f(n) - \int_1^n f(x)\,dx \leqq f(1)$$

or $$\sum_1^{n-1} u_n - \int_1^n f(x)\,dx \leqq f(1) \qquad \text{(for all } n\text{)}$$

and therefore

$$\lim_{n\to\infty} \left\{ \sum_1^{n-1} u_n - \int_1^n f(x)\,dx \right\} \text{ exists}$$

Now $\int_1^x f(x)\,dx$ is a positive steadily increasing function of x but we have proved that the expression

$$u_1+u_2+u_3+ \dots +u_{n-1} - \int_1^n f(x)\,dx$$

tends to a limit. Hence the series

$$\sum u_n$$

tends to a limit or to infinity according as the integral

$$\int_1^n f(x)\,dx$$

tends to a limit or to infinity. This proves the theorem.

Denote $$\int_1^n f(x)\,dx \text{ by } F(n)$$

If $\lim_{n\to\infty} F(n)$ exists let it be l.

Then we may deduce that

$$\sum_1^\infty u_n \leqq f(1)+l$$

As a simple example,

$$\sum_1^\infty \frac{1}{n^2+1} \leqq \frac{1}{2}+\lim_{n\to\infty}\left\{\int_1^n \frac{dx}{1+x^2}\right\}$$

$$= \frac{1}{2}+\lim_{n\to\infty}\Big[\text{arc tan } x\Big]_1^n$$

$$= \frac{1}{2}+\frac{\pi}{4}$$

We are now in a position to extend our comparison tests for convergence. Consider the series

$$S = \sum n^{-s} \qquad 1^{-s}+2^{-s}+3^{-s}+4^{-s}+\ \ldots$$

where s is rational.

This series is known to diverge if $s = 1$ and it is clearly divergent if $s = 0$ or if $s < 0$.

If $s > 0$ then u_n is a decreasing function of n and the test above may be applied. Here

$$F(n) = \int_1^n \frac{dx}{x^s} = \frac{n^{1-s}-1}{1-s} \quad \text{provided} \quad s \neq 1$$

Now if $\qquad s > 1, \qquad \lim_{n\to\infty} F(n) = \dfrac{1}{s-1}$

if $\qquad s < 1, \qquad \lim_{n\to\infty} F(n) \to \infty$

Hence we have the result that the series $\sum_1^\infty n^{-s}$ converges if $s > 1$ and diverges if $s \leqq 1$.

If it converges the sum is less than $s/(s-1)$.

Series which Contain both Positive and Negative Terms

Our previous discussion was restricted to series all of whose terms were positive. Naturally if all terms are negative the same arguments apply. Furthermore if the condition that all terms are positive, or negative, is violated only for a finite number of terms, we may apply our results to the remainder of the series.

It may happen, however, that the changes of sign persist throughout the series, and if so we must reconsider the arguments.

We have a further complication now, since there is no longer a simple distinction between convergence and divergence: the series may oscillate. We must now consider series whose terms are both positive and negative.

Absolute Convergence

Let
$$S = u_1 + u_2 + u_3 + \dots$$
be a series containing both positive and negative terms. If the associated series
$$S' = |u_1| + |u_2| + |u_3| + \dots$$
consisting of the absolute values of u_1, u_2, u_3, etc., is convergent, then the original series is said to be absolutely convergent. The converse is NOT true. E.g. the series
$$1 - \tfrac{1}{2} + \tfrac{1}{3} - \tfrac{1}{4} + \tfrac{1}{5} - \dots$$
is convergent, but the series of absolute values, viz.
$$1 + \tfrac{1}{2} + \tfrac{1}{3} + \tfrac{1}{4} + \dots$$
is divergent.

It is easy to show by the results for rearranging terms, that if a series is absolutely convergent, then the series formed by the positive terms only or the negative terms only are also convergent, and that the sum of the series is equal to the sum of the positive terms less the sum of the negative terms.

Conditionally Convergent Series

If the series

$\sum u_n$ converges, but the series

$\sum |u_n|$ diverges, then the series $\sum u_n$ is said

to be conditionally convergent.

In a conditionally convergent series the sums of the positive or negative terms taken separately are divergent.

Further, we cannot rearrange the terms in a conditionally convergent series without altering the sum, although we can do this with an absolutely convergent series.

Convergence Tests for Conditionally Convergent Series

The first difficulty we meet is that there are no comparison tests for conditionally convergent series.

There frequently occur, however, series in which the terms are alternately positive and negative. Such series are called Alternating Series. In an important special case of this type of series we can give a test.

If $u(n)$ is a positive function of n which tends steadily to zero, then the series

$$U = u(0) - u(1) + u(2) - u(3) \dots$$

is convergent and its sum U satisfies the relation

$$u(0) < U < [u(0) - u(1)]$$

Proof. The sum U_n of $n+1$ terms is

$$u(0) - u(1) + u(2) + \dots + (-1)^n u(n)$$

Then $\quad U_{2n+1} - U_{2n-1} = u(2n) - u(2n+1) \geqq 0$

and $\quad U_{2n} - U_{2n-2} = -[u(2n-1) - u(2n)] \leqq 0$

Hence the sequence U_0, U_2, U_4, U_8 is a decreasing sequence and therefore tends to a limit or to $-\infty$, and also the sequence U_1, U_3, U_5, U_7 is an increasing sequence and so tends to a limit or to infinity.

But $\lim_{n\to\infty} (U_{2n+1}-U_{2n}) = \lim_{n\to\infty} \{(-1)^{2n+1}u(2n+1)\} = 0$

But the above sequences tend to limits and so these limits must be equal. But the sum

$$U_{2n+1} < U_1 \quad \text{i.e.} \quad > u(0)-u(1)$$

and the sum $$U_{2n} < u(0)$$

and since both tend to the same limit, the sum of the complete series U must lie between $u(0)$ and $u(0)-u(1)$.

We shall end our discussion of conditionally convergent series with the statement that a conditionally convergent series remains convergent if we multiply its terms by any sequence of positive but decreasing factors, i.e. if

$$\alpha_1, \alpha_2, \alpha_3 \ldots$$

is a series of positive decreasing functions of n, and if the series $\sum u_n$ is convergent, then so is the series $\sum \alpha_n u_n$.

Series whose Terms are Functions of a Continuous Variable

So far, we have considered series each term of which was a function of a positive integer n.

Consider now a series

$$S = u_1(x)+u_2(x)+u_3(x)+ \ldots$$

We must consider whether the sum of such a series, assuming that it is convergent, represents a continuous function. We must also determine the conditions under which it is legitimate to integrate or differentiate such series. For example, even if the series converges to $S(x)$ it does not necessarily follow that

$$\frac{dS}{dx} = u_1'(x)+u_2'(x)+u_3'(x) + \ldots$$

where
$$u_n'(x) = \frac{du_n}{dx}$$

A particular example of series of functions of x which is of great usefulness is a series in which $u_n(x)$ is a power of x with a constant coefficient.

Such a series is known as a *Power Series*, e.g.

$$S = a_0+a_1x+a_2x^2+a_3x^3+ \ldots + a_nx^n + \ldots \quad (6.3)$$

Such series can arise as the solution of differential equations.

Theorem.

If
$$\lim_{n\to\infty}\left|\frac{a_{n+1}}{a_n}\right| = \frac{1}{R}$$

then the power series (6.3) converges absolutely when $x < R$, diverges when $x > R$. We can state no conclusions when $x = R$.

Proof. This follows immediately from the theorem on the ratio test on p. 125.

Theorem. If, when $x = R$, none of the terms in the series (6.3) exceeds numerically a finite quantity α then the series will be absolutely convergent for x numerically less than R.

Proof. We can write the series as

$$S = a_0+a_1R\left(\frac{x}{R}\right)+a_2R^2\left(\frac{x}{R}\right)^2 + \ldots$$

then the terms of this series are numerically less than the terms of the series

$$S_1 = \alpha+\alpha\left(\frac{x}{R}\right)+\alpha\left(\frac{x}{R}\right)^2+\alpha\left(\frac{x}{R}\right)^3+\ldots$$

The series converges absolutely so long as $x < R$.

Definition. If a series whose terms are functions of x is convergent for values of x such that $a < x < b$, then the interval (a, b) is called the *interval* of convergence.

Uniform Convergence

Let $\qquad S = u_1(x)+u_2(x)+\ \ldots\ +u_n(x)+R_n(x)$

Then a condition for convergence is that $R_n(x) \to 0$ as $n \to \infty$. In our earlier sections we made this precise by saying that this meant that we could choose a number n_0 such that for

$$n \geqq n_0,$$

$$|R_n(x)| < \varepsilon$$

The number n_0 will, however, now depend on the value of x chosen. This consideration leads us to a very important definition.

Definition. A series

$$S = u_1(x)+u_2(x)+u_3(x)+\ \ldots\ +u_n(x)+R_n(x)$$

is said to converge *uniformly* in an interval if it is possible to choose a number n_0 such that for $n \geqq n_0$, *and for all values of* x *in the interval*, the quantity $R_n(x)$ shall be less than any given quantity ε, however small.

We are now able to discuss some questions of continuity of series.

Theorem. If the series

$$S(x) = u_1(x)+u_2(x)+u_3(x)+\ \ldots$$

converges uniformly in the interval $a \leqq x \leqq b$ and if each term is a continuous function in the interval, then the sum $S(x)$ is also a continuous function in the interval.

Proof. Let x_1 and x_2 be two values of x in the interval. All we need do is to show that if x_1 is sufficiently near to x_2 then

$$|S(x_1)-S(x_2)| < \varepsilon \quad \text{where } \varepsilon \text{ is given}$$

But $\quad S(x_1)-S(x_2) = S_n(x_1)-S_n(x_2)+R_n(x_1)-R_n(x_2)$

Now since the series is uniformly convergent we can choose n_0 so that for $n \geqq n_0$ both $R_n(x_1)$ and $R_n(x_2)$ are $< \varepsilon/3$. Further, since $S_n(x)$ is the sum of a finite number of continuous functions we can evidently take x_1 so near to x_2 that

$$|S_n(x_1)-S_n(x_2)| < \varepsilon/3$$

Considering this we see that

$$|S(x_1)-S(x_2)| < \varepsilon$$

which proves the theorem for x within the interval.

The proof for either end of the interval follows by taking $x_1 = a$, then $x_1 = b$.

An immediate consequence is that a power series

$$a_0+a_1x+a_2x^2+ \ldots$$

represents a continuous function within its interval of convergence $(-R, R)$. It is only necessary to show that if $a \leqq x \leqq b$ then the series is uniformly convergent. Note that the interval (a, b) is within the interval of convergence $(-R, R)$.

Proof. Take $\rho < R$ but $> |a|$ and $> |b|$, then the series $\sum a_n\rho^n$ is absolutely convergent but

$$a_nx^n = a_n\rho^n\left(\frac{x}{\rho}\right)^n$$

$$\therefore \quad |a_nx^n| < |a_n\rho^n|$$

Let $r_n(x)$ be the remainder after n terms of the series $\sum a_n x^n$

$$\therefore \quad r_n(x) < |a_n\rho^n| + |a_{n+1}\rho^{n+1}| + \ldots$$

But we can choose N so that for $n \geqq N$ the sum

$$|a_n\rho^n| + |a_n\rho^{n+1}| + \ldots < \varepsilon$$

and for this value of N we have

$$r_n(x) < \varepsilon \quad (n \geqslant N)$$

for all x in the interval (a, b).

This is the condition for uniform convergence of the Power Series and hence proves its continuity.

By employing similar arguments the student will be able to prove that if the terms of a series are continuous functions of x in an interval $a \leqq x \leqq b$ and if they are numerically less than the terms of an absolutely convergent series, then the series is uniformly convergent in the interval (a, b).

Differentiation and Integration of Series

We already have rules for differentiating and integrating a sum of terms but the rules apply for finite sums only. It requires justification if we are to apply them to the sum of an infinite number of terms.

It is important to obtain this extension to the rules because many functions are easiest to investigate in the form of an infinite series and some are defined in this form.

Let us say at once that we shall proceed to prove that a series $\sum U_n(x)$ which is *uniformly* convergent may be integrated term by term and the resulting series is convergent and gives an expression for the integral.

The case for differentiation is rather different. If a series $\sum U_n(x)$ converges then we may differentiate term by term to get the derivative provided that the *resulting series* $\sum U_n'(x)$ is *uniformly* convergent.

Theorem I. Let

$$f(x) = u_1(x)+u_2(x)+u_3(x)+ \dots$$

be a series which is uniformly convergent for $a \leqq x \leqq b$, then the series

$$\int_c^x u_1(x)\,dx+\int_c^x u_2(x)\,dx+\int_c^x u_3(x)\,dx+ \dots$$

where $$a \leqq c < x \leqq b$$

is convergent and its sum is $\int_c^x f(x)\,dx$

As usual let ε be any given small quantity. Let

$$f(x) = S_n(x)+R_n(x)$$

and define

$$\sigma_n(x) = \int_c^x S_n(x)\,dx \quad \text{and} \quad \rho_n(x) = \int_c^x R_n(x)\,dx$$

So that $$\int_c^x f(x)\,dx = \sigma_n(x)+\rho_n(x)$$

Now, because of the uniform convergence of the series we can choose a number N so that for $n \geqq N$, $R_n(x) < \varepsilon$ for all values of x in the interval (a, b).

If we do this then

$$\rho_n(x) = \int_c^x R_n(x)\,dx < \varepsilon(x-c)$$

This means that for $n \geqq N$

$$\int_c^x f(x)\,dx-\sigma_n(x) < \varepsilon(x-c)$$

This statement is what we mean by saying that

$$\lim_{n\to\infty} \left\{\int_c^x f(x)\,dx-\sigma_n(x)\right\} = 0$$

i.e.

$$\int_c^x f(x)\,dx = \lim_{n\to\infty} \sigma_n(x) = \int_c^x u_1(x)\,dx + \int_c^x u_2(x)\,dx + \ldots$$

which proves the theorem.

Theorem II. Let $$u_1(x)+u_2(x)+u_3(x)+ \ldots$$

be a convergent series for $a \leqq x \leqq b$ and let its sum be $f(x)$, Then if the series formed by differentiating each term, viz.

$$u_1'(x)+u_2'(x)+u_3'(x)+ \ldots$$

is uniformly convergent, in the interval (a, b), its sum will be equal to $f'(x)$.

Proof. Let $$F(x) = u_1'(x)+u_2'(x)+ \ldots$$

Since this series is uniformly convergent we may use Theorem I.

$$\text{So} \quad \int_c^x F(x)\,dx = \int_c^x u_1'(x)\,dx + \int_c^x u_2'(x)\,dx + \ldots$$

$$= (u_1(x)-u_1(c))+(u_2(x)-u_2(c))+ \ldots$$

$$= u_1(x)+u_2(x)+ \ldots -(u_1(c))+u_2(c) \ldots$$

$$= f(x)+\text{const.}$$

$$\therefore \quad F(x) = f'(x)$$

which proves the theorem.

Changing the Sum of a Conditionally Convergent Series by Rearranging Terms

We are now in a position to prove the statement in the introduction on p. 2.

We said that while the series

$$S = 1-\tfrac{1}{2}+\tfrac{1}{3}-\tfrac{1}{4}+\tfrac{1}{5} + \ldots = \log_e 2 \tag{6.4}$$

the series in which we took two positive terms and then one negative term, i.e.

$$S' = 1+\tfrac{1}{3}-\tfrac{1}{2}+\tfrac{1}{5}+\tfrac{1}{7}-\tfrac{1}{4}+\ \ldots = \tfrac{3}{2}\log_e 2 \qquad (6.5)$$

Take the sum of $3n$ terms of (6.5),

$$S'_{3n} = 1+\frac{1}{3}+\frac{1}{5}+\ \ldots\ +\frac{1}{4n-1}-\frac{1}{2}-\frac{1}{4}-\ \ldots\ -\frac{1}{2n}$$

$$= S_{2n}+\frac{1}{2n+1}+\frac{1}{2n+3}+\ \ldots\ +\frac{1}{4n-1} \qquad (6.6)$$

where S_{2n} is the sum of $2n$ terms of 6.4. Now let us study the terms other than S_{2n} in (6.6). If we add and subtract the sequence

$$\frac{1}{2n+2}+\frac{1}{2n+4}+\ \ldots\ +\frac{1}{4n}$$

we have

$$S'_{3n} = S_{2n}+\left[\frac{1}{2n+1}-\frac{1}{2n+2}+\frac{1}{2n+3}-\frac{1}{2n+4}\ \ldots\right.$$

$$\left.+\frac{1}{4n-1}-\frac{1}{4n}\right]+\left(\frac{1}{2n+2}+\frac{1}{2n+4}\ \ldots\ +\frac{1}{4n}\right)$$

The term in square brackets is less than $\dfrac{n}{(2n+1)(2n+2)}$ as is shown by taking positive and negative terms in pairs and adding. Hence when n tends to infinity this part of the expression vanishes. The term in round brackets is

$$\frac{1}{2}\left(\frac{1}{n+1}+\frac{1}{n+2}+\ \ldots\ +\frac{1}{2n}\right) = \frac{1}{2}\frac{1}{n}\sum_{r=1}^{n}\left(\frac{1}{1+r/n}\right)$$

and $$\lim_{n\to\infty}\frac{1}{n}\sum_{r=1}^{n}\left(\frac{1}{1+r/n}\right) = \int_1^2 \frac{dx}{x} = \log_e 2$$

Hence $$\lim S'_{3n} = \lim S_{2n}+\tfrac{1}{2}\log_e 2$$

i.e.
$$S' = S+\tfrac{1}{2}\log_e 2$$
$$= \tfrac{3}{2}\log_e 2$$

It will be a useful exercise for the reader to show similarly that the sum

$$S'' = 1-\tfrac{1}{2}-\tfrac{1}{4}+\tfrac{1}{3}-\tfrac{1}{6}-\tfrac{1}{8}+\tfrac{1}{5} - \ldots = \tfrac{1}{2}\log_e 2$$

Infinite Integrals

Infinite Integrals are generally classified under two headings referred to as infinite integrals of the first and second kind.

Infinite Integrals of the First Kind

These are of the form

$$\int_a^{\infty} f(x)\,dx \quad \text{or} \quad \int_{-\infty}^{\infty} g(x)\,dx$$

where the infinity symbol occurs in the limits of integration. It should be noted at once that the result or value of the integration may be finite just as the sum of a convergent but infinite series is finite. It is fairly simple to frame a definition.

If the function $f(x)$ is continuous for all $x \geqq a$, and if

$$\lim_{m\to\infty} \int_a^m f(x)\,dx = l$$

then we say that

$$\int_a^{\infty} f(x)\,dx \quad \text{is convergent and has the value } l$$

If $\lim_{m\to\infty} \int_a^m f(x)\,dx$ tends to infinity we say that

$$\int_a^{\infty} f(x)\,dx \quad \text{diverges to infinity}$$

If $\lim_{m \to \infty} \int_a^m f(x)\, dx$ does neither of these things then the integral is said to oscillate. It may do so finitely or infinitely.

Infinite Integrals of the Second Kind

These are integrals in which the integrand is not continuous at some point or points in the range of integration. We need consider the case of one point only and we can assume, with no loss of generality, that this point is the lower limit of integration. We can now give a definition. Let $f(x)$ be continuous except when $x = a$, but $f(x) \to \infty$ when $x \to a$ by positive values. Then if

$$\lim_{\varepsilon \to +0} \int_{a+\varepsilon}^{b} f(x)\, dx \quad \text{has a value } l$$

we say that the integral

$$\int_a^b f(x)\, dx \quad \text{is convergent and has the value } l$$

If we imagine the non-continuity of $f(x)$ to be associated with the upper limit we define the integral as

$$\lim_{\varepsilon \to +0} \int_a^{b-\varepsilon} f(x)\, dx$$

Assuming that $f(x)$ is positive, a condition for convergence is that

$$\int_{a+\varepsilon}^{b} f(x)\, dx < K \quad \text{for all values of } \varepsilon, \text{ however small} \quad (6.7)$$

This follows since the integral is an increasing function as ε decreases. If the condition (6.7) is not satisfied then the integral diverges to infinity.

Examples.

$$\int_a^b \frac{dx}{\sqrt{(x-a)(x-b)}} \quad \text{is convergent}$$

$$\int_0^\pi \frac{dx}{\sqrt{(\sin x)}} \quad \text{is convergent}$$

The reader should note that an infinite integral of this kind is essentially a double limit, being the limit of the limit of a sum. It is thus a complex type of function and should be studied carefully.

There are a number of elementary points to note at once.

(1) If $f(x)$ is always positive then the integral is a steadily increasing function and so must tend to a limit or $+\infty$.

(2) In such a case a necessary and sufficient condition for convergence is that

$$\int_a^x f(x)\, dx \leqq K$$

for all $x > a$.

A useful comparison test is that with the integral

$$\int_a^\infty \frac{dx}{x^s} \quad \text{for} \quad a > 0$$

This we have already studied.

The conclusions are:

If $\quad f(x) < Kx^{-s} \quad$ where $\quad S > 1 \quad$ then

$$\int_a^x f(x)\, dx \quad \text{is convergent for } x > a$$

but if $f(x) > Kx^{-s}$ where $S \leqq 1$ the integral is divergent for $x > a$.

Examples

$\int_a^\infty \frac{dx}{x^{1/3}}$ is divergent $\qquad \int_1^\infty \frac{dx}{x^{4/3}}$ is convergent

$\int_a^\infty \frac{x\,dx}{a^2+x^2}$ is divergent $\qquad \int_a^\infty \frac{x^2\,dx}{a^2+x^2}$ is convergent

State whether the series of positive terms whose *n*th term is shown below are convergent or divergent.

1. $\frac{n}{(n+1)^2}$ $\qquad$ 2. $\frac{n^3}{n!}$

3. $\frac{1}{(2n+3)^s}$ $\qquad$ 4. $\frac{1}{n(n+1)}$

5. Show that the series $1+\frac{1}{2}-\frac{1}{3}+\frac{1}{4}-\frac{1}{5}+\frac{1}{6}\ldots$ is divergent.

6. Show that $\frac{3}{2} < \sum_1^\infty \frac{1}{n^2} < 2$

7. Show that the series

$$1 + x + \frac{x^2}{2!} + \frac{x^3}{3!} + \ldots + \frac{x^n}{n!}$$

is uniformly convergent.

CHAPTER SEVEN

ADDITIONAL THEOREMS IN DIFFERENTIAL AND INTEGRAL CALCULUS

The Second Mean Value Theorem

We already know that if $f(x)$ has a derivative $f'(x)$ in (a, b) then

$$f(b)-f(a) = (b-a)f'(x_1) \quad \text{where} \quad a < x_1 < b$$

We shall now extend this theorem.

Suppose $f(x)$ has also a second derivative $f''(x)$ in (a, b), a supposition, be it noted, which implies the continuity of $f'(x)$ in the interval.

Then the Second Mean Value Theorem states that

$$f(b)-f(a) = (b-a)f'(a)+\tfrac{1}{2}(b-a)^2 f''(x_2) \qquad a < x_2 < b \quad (7.1)$$

The proof follows the lines of that on p. 62.

We consider the function

$$F(x) = f(b)-f(x)-(b-x)f'(x) - \left(\frac{b-x}{b-a}\right)^2 \{f(b)-f(a)-(b-a)f'(a)\}$$

This function vanishes for $x = a$ and $x = b$ and its derivative is

$$F'(x) = \frac{2(b-x)}{(b-a)^2}\{f(b)-f(a)-(b-a)f'(a)-\tfrac{1}{2}(b-a)^2 f''(x)\}$$

This must vanish for some value x_2 such that $a < x_2 < b$ which gives at once

$$f(b)-f(a) = (b-a)f'(a)+\tfrac{1}{2}(b-a)^2 f''(x_2)$$

Taylor's Formula

This is a further generalisation of the two Mean Value Theorems.

If $f(x)$ has derivatives of the first n orders in (a, b) then

$$f(b)-f(a) = (b-a)f'(a)+\tfrac{1}{2}(b-a)^2f''(a)+\frac{(b-a)^3}{1\cdot 2.3}f'''(a)\ldots$$

$$+\frac{(b-a)^{n-1}}{(n-1)!}f^{n-1}(a)+\frac{(b-a)^n}{n!}f^n(x_n) \quad (7.2)$$

where $a < x_n < b$.

The proof proceeds exactly on the lines already given. We form the function

$$F_n(x) = f(b)-f(x)-(b-x)f'(x)-\frac{(b-x)^2}{2}f''(x)$$

$$-\frac{(b-x)^{n-1}}{(n-1)!}f^{n-1}(x)$$

and study the behaviour for $x = a$ and $x = b$ of the function

$$F_n(x)-\left(\frac{b-x}{b-a}\right)^n F_n(a)$$

This vanishes for $x = a$ and $x = b$ and has the derivative

$$\frac{n(b-x)^{n-1}}{(b-a)^n}\left\{F_n(a)-\frac{(b-a)^n}{n!}f^n(x)\right\}$$

This vanishes for x_n where $a < x_n < b$ and the result follows. This formula is often expressed in the equivalent form

$$f(x+h) = f(x)+hf'(x)+\frac{h^2}{2}f''(x)+\ \ldots\ +\frac{h^{n-1}}{(n-1)!}f^{n-1}(x)$$

$$+\frac{h^n}{n!}f^n(x+\theta h)$$

where $0 < \theta < 1$.

An important special case where $x = 0$ is known as MacLaurin's Theorem:

$$f(h) = f(0)+hf'(0)+ \dots + \frac{h^n}{n!} f^n(\theta h) \qquad (7.3)$$

Taylor's and MacLaurin's Series

If $f(x)$ and all its derivatives are continuous for the range (a, b) and if

$$\lim_{n\to\infty} \frac{(x-a)^n}{n!} f^n[a+\theta(x-a)] = 0$$

then the infinite series for $f(x)$, viz.

$$f(x) = f(a)+(x-a)f'(a)+ \dots$$

converges to the value $f(x)$.

This is known as *Taylor's Series*. Note that it is not only necessary for $f(x)$ to have derivatives of all values, but it is essential to prove that the 'remainder' in Taylor's Formula tends to zero when n tends to infinity.

The corresponding series, called MacLaurin's Series, leads to

$$f(x) = f(0)+xf'(0)+\frac{x^2}{2} f''(0)+ \dots$$

with, of course, the same conditions for convergence.

Example. Let $$f(x) = \sin x$$

$$f'(x) = \cos x \qquad f''(x) = -\sin x \quad \text{etc.}$$

The remainder is $\dfrac{x^n}{n!} f^n(\theta x) = \dfrac{x^n}{n!} \sin\left(\theta x+\dfrac{n\pi}{2}\right)$

This cannot be greater than $x^n/n!$ which tends to zero (p. 145). Hence

$$\sin x = x-\frac{x^3}{3!}+\frac{x^5}{5!}-\frac{x^7}{7!}+ \dots$$

Application to the Theory of Equations

If $f(x)$ is a rational integral function of degree n, then it is known from algebra that the equation

$$f(x) = 0$$

has n roots. These roots are not necessarily all different nor need they be all real. The values are referred to as the zeros of $f(x)$. Where these are multiple roots let

$$f(x) = (x-\alpha)^r g(x)$$

Then it is obvious that the first $(r-1)$ derivatives of $f(x)$ will contain $(x-\alpha)$ as a factor but the rth derivative will not. Hence any multiple root of $f(x) = 0$ is a root of $f'(x) = 0$ and we may obtain the multiple roots by determining the Greatest Common Multiple of $f(x)$ and $f'(x)$.

Roots of the Equation $f(x) = 0$ when $f(x)$ is a Continuous Function

Let $f(x)$ be *any* continuous function. We are already in possession (pp. 27, 57) of information which enables us to draw some valuable conclusions.

(1) If $f(a)$ and $f(b)$ have opposite signs then there is *at least one* root of $f(x) = 0$ in the interval (a, b).
(2) If $f'(x)$ is continuous and does not vanish in (a, b) then $f(x)$ is either an increasing or a decreasing function throughout the interval (a, b), and in this case there is only *one root* in the interval (a, b).
(3) If $f(x)$ and $f'(x)$ are continuous, then between two consecutive roots of $f(x) = 0$ there is a root of $f'(x) = 0$. (This is just Rolle's Theorem.)

Newton's Method of Approximation

Suppose that $f(x)$, $f'(x)$ and $f''(x)$ are continuous and that $f(x)$ and $f'(x)$ do not vanish for the same value of x. This last proviso is, of course, the condition that the equation $f(x)$ does not have a multiple root for the range of values of x under consideration.

Let α be an approximation to the root of $f(x) = 0$ whose true value is $\alpha+h$.

Then by the Second Mean Value Theorem

$$0 = f(\alpha+h) = f(\alpha)+hf'(\alpha)+\tfrac{1}{2}h^2f''(\alpha+\theta_2 h)$$

i.e.
$$h = \frac{-f(\alpha)}{f'(\alpha)}-\tfrac{1}{2}h^2\frac{f''(\alpha+\theta_2 h)}{f'(\alpha)}$$

Now if h is small, which will be so if α is a reasonable approximation, and if we may neglect h^2 in comparison with h, we get a *better approximation* to the root by taking it to be

$$\alpha_1 = \alpha+h_1 \quad \text{or} \quad \alpha_1 = \alpha-\frac{f(\alpha)}{f'(\alpha)}$$

Note that we have *not* found the true value of h but we have found h_1 which is an approximate value of h.

We could, of course, use the new value α_1 as our basic approximation and proceed as before.

Some caution is necessary in using this method. However, before discussing this let us give an example.

Example. Find an approximate root of the equation

$$x^2 = 5$$

Take $x = 2$ as an approximate answer. Then a better answer is

$$x_1 = 2+\tfrac{1}{4} = 2{\cdot}25$$

We can repeat this,

$$x_2 = 2{\cdot}250-0{\cdot}07/4{\cdot}5$$
$$= 2{\cdot}235$$

Four figure tables give 2·236, so we are getting very close.

Degree of Approximation

We have not *proved* that α_1 is closer to the actual root of $f(x) = 0$ than α is. Further, we do not know the *degree* of approximation. It is worth while to examine the problem more closely.

In an interval (a, b) let $f(a)$ and $f(b)$ have opposite signs and suppose that in the interval neither $f'(x)$ nor $f''(x)$ vanish. Let a be the end of the interval for which $f(a)$ and $f''(a)$ have the same sign. This is shown in Figure 21.

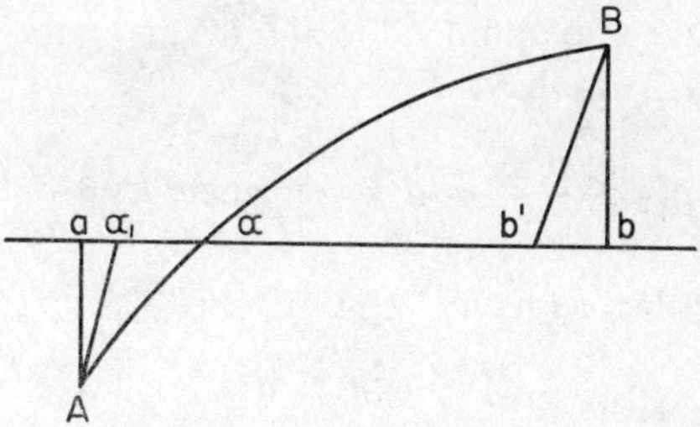

FIG. 21

The equation of the tangent at A is

$$y = f(a)+(x-a)f'(a)$$

but when, on this line,

$$y = 0, \qquad x = \alpha_1$$

so

$$\alpha_1 = a-f(a)/f'(a)$$

The condition on the equal signs of $f(a)$ and $f''(a)$ ensures that the point α_1 lies between a and α. If this condition is not satisfied α_1 could be further from α than a and our process will fail.

We wish now to try to estimate the degree of approxima-tion. Draw a line from B parallel to the tangent at A and let this cut the Y axis at b'. Then the equation of this line is

$$y = f(b)+(x-b)f'(a)$$

so
$$b' = b-\frac{f(b)}{f'(a)}$$

so
$$(b'-\alpha_1) = -\tfrac{1}{2}(b-a)^2\frac{f''(x_1)}{f'(a)}$$

where x_1 is some value of x in (a, b).

Let
$$d = b-a, \qquad d_1 = b'-\alpha_1$$

Then if G is the maximum value of $f''(x)$ and g the least value of $f'(x)$ in (a, b) we have

$$d_1 \leqq \frac{d^2G}{2g} = d^2k \qquad \text{where } k = \frac{G}{2g}$$

If we repeat this we have

$$d_2 \leqq d_1^2k \leqq d^4k^3$$

Now the error in taking α_1 instead of α as the root is certainly less than $b'-\alpha_1$, i.e. less than d^2k. Also if dk is less than one the sequence

$$d^2k, \quad d^4k^3, \ldots$$

converges to zero very rapidly, and so our approach to the true root α becomes very rapid also.

Envelopes

Consider the equation

$$f(x, y, \alpha) = 0$$

where α is a constant. This represents a curve. Suppose we allow α to take a succession of values $\alpha_1, \alpha_2 \ldots$ then we will obtain a succession of curves, each one corresponding to a different value of α. We have, if we like, a *family* of curves.

A constant which can, so to speak, vary among the family is called a *parameter*.

Different members of the family will, in general, intersect. Consider the two members defined by

$$f(x, y, \alpha) = 0 \qquad (7.4)$$

$$f(x, y, \alpha+\delta\alpha) = 0 \qquad (7.5)$$

These members are adjacent to each other if $\delta\alpha$ is small. The coordinates of their point, or points, of intersection are given by solving the equations (7.4) and (7.5), or the equivalent equations

$$f(x, y, \alpha) = 0$$

$$\frac{f(x, y, \alpha+\delta\alpha)-f(x, y, \alpha)}{\delta\alpha} = 0$$

Now if we proceed to the limit when $\delta\alpha$ tends to zero, we arrive at the equations for intersection

$$f(x, y, \alpha) = 0$$

$$\frac{\partial}{\partial\alpha}\{f(x, y, \alpha)\} = 0 \qquad (7.6)$$

These two equations are, in effect, equations expressing both x and y in terms of a parameter α. If we think of α as varying and eliminate it between these two equations we get a solution

$$\phi(x, y) = 0 \qquad (7.7)$$

This curve is called the Envelope of the family of curves

$$f(x, y, \alpha) = 0 \qquad (7.8)$$

The envelope touches every member of the family at some point. For the gradient at x, y on (7.8) is given by

$$\frac{\partial f}{\partial x}+\frac{\partial f}{\partial y}\frac{dy}{dx} = 0 \qquad (7.9)$$

But the process of eliminating α between (7.4) and (7.6) really means that (7.8) is the equation of the envelope provided we solve (7.6) for α in terms of x and y and substitute in (7.8). Doing this and now regarding α as $\alpha(x, y)$ we have the gradient of the envelope given by

$$\frac{\partial f}{\partial x}+\frac{\partial f}{\partial y}\frac{dy}{dx}+\frac{\partial f}{\partial \alpha}\frac{d\alpha}{dx} = 0 \tag{7.10}$$

Now suppose the point (x, y) is on *both* the curve and the envelope, then for this value of xy

$$\frac{\partial f}{\partial \alpha} = 0$$

and hence for this value of xy (7.9) and (7.10) are identical. This shows that for a point on both the curve and the envelope dy/dx is the same for both.

Of course (7.9) depends on α but there is a particular pair of values of x and y for which dy/dx calculated from (7.8) is identical with the value of dy/dx calcultated from (7.10) which does not involve α explicitly.

Example.

$$y = mx+\frac{a}{m} \tag{7.11}$$

so

$$f(x, y, m) = -y+mx+\frac{a}{m}$$

$$\frac{\partial f}{\partial m} = x-\frac{a}{m^2}$$

Eliminating m we get

$$y^2 = 4ax$$

This parabola is the envelope of the family of straight lines given by (7.11).

Indeterminate Forms

As a final example of the use of differential calculus we shall discuss the interpretation of a function $f(x)$ which for some value of its argument takes the form $0/0$, ∞/∞, etc. Suppose this occurs when $x = a$.

Strictly speaking, of course, such forms have absolutely no meaning whatever.

However, there may well be a limit of $f(x)$ when $x \to a$ and this opens up the possibility of *defining* $f(x)$ for $x = a$. We emphasize that this definition is an arbitrary one. It has the advantage of making $f(x)$ continuous for $x = a$.

Sometimes we can find the limit by direct methods, e.g. $(1+x)^{1/x}$ takes the indeterminate form when $x = 0$.

However, we already know that $\lim_{x\to 0}\{(1+x)^{\frac{1}{x}}\} = e$

Take again $$x \log x$$
This assumes the form

$$0 \times \infty \quad \text{for} \quad x = 0$$

but we know that $\lim_{x\to 0}\{x \log x\} = 0$

$\dfrac{\text{Sin}\, x}{x}$ takes the form $\dfrac{0}{0}$ and we know the limit for $x \to 0$ is 1.

When these methods are not available we can use the calculus.

Theorem. If $\phi(a)$ and $\psi(a)$ are both zero or both infinite

then $$\lim_{x\to a}\frac{\phi(x)}{\psi(x)} = \lim_{x\to a}\frac{\phi'(x)}{\psi'(x)}$$

Proof. We require a slight generalisation of the Mean Value Theorem. Viz. if $\phi(x)$, $\phi'(x)$, $\psi(x)$, $\psi'(x)$ are all continuous in the interval $a \leqq x \leqq b$, and if $\psi'(x)$ is not zero in the interval, then

$$\frac{\phi(b)-\phi(a)}{\psi(b)-\psi(a)} = \frac{\phi'(x_1)}{\psi'(x_1)}$$

where $a < x_1 < b$.

We can prove this by applying Rolle's Theorem to

$$F(x) = \frac{\phi(b)-\phi(a)}{\psi(b)-\psi(a)}\{\psi(x)-\psi(a)\}-\{\phi(x)-\phi(a)\}$$

since $F(a) = 0 = F(b)$ and we may divide by $\psi'(x_1)$ since this, by definition, is not zero. We now suppose that $\phi(a) = \psi(a) = 0$ and put $b = x$, then

$$\frac{\phi(x)}{\psi(x)} = \frac{\phi'(x_1)}{\psi'(x_1)} \quad \text{where } a < x_1 < x$$

Then $$\lim_{x\to a}\frac{\phi(x)}{\psi(x)} = \lim_{x\to a}\frac{\phi'(x_1)}{\psi'(x_1)} = \lim_{x\to a}\frac{\phi'(x)}{\psi'(x)}$$

Now suppose that $$\lim_{x\to\infty}\frac{\phi'(x)}{\psi'(x)} = l$$

but that $\lim_{x\to\infty}\phi(x) \to \infty$ and also $\lim_{x\to\infty}\psi(x) \to \infty$

Then $$\lim_{x\to\infty}\frac{\phi(x)}{\psi(x)} = \lim_{x\to\infty}\frac{\phi'(x)}{\psi'(x)} = l$$

Proof. Let C be a large but finite value of x, then consider the identity

$$\frac{\phi(x)}{\psi(x)} = \frac{\phi'(x_1)}{\psi'(x_1)} \times \frac{1-\dfrac{\psi(C)}{\psi(x)}}{1-\dfrac{\phi(C)}{\phi(x)}} \quad \text{when } C < x_1 < x$$

Now since $$\lim_{x\to\infty} \frac{\phi'(x_1)}{\psi'(x_1)} = l$$

we can choose a value of C, say $C = M$, so that for $x_1 > M$ the ratio

$$\frac{\phi'(x)}{\psi'(x)} < |l+\varepsilon|$$

where ε is as small as we wish.

Having established this we may, regarding C as fixed but allowing x to increase still further, choose a value of x such that

$$\frac{1-\dfrac{\psi(C)}{\psi(x)}}{1+\dfrac{\phi(C)}{\phi(x)}} < |1+y|$$

where y is arbitrarily small.

This is possible since $\phi(x)$ and $\psi(x)$ tend to infinity with x. This now means that we can make

$$\frac{\phi(x)}{\psi(x)} < l\left|1+\frac{\varepsilon}{l}\right||1+y|$$

i.e. differ from l by as little as we please. This proves the theorem.

If $$\phi(x) \to \infty \quad \text{and} \quad \psi(x) \to \infty \quad \text{as } x \to a$$

by putting $$x = a+\frac{1}{z}$$

we reduce the problem to the previous one.

Example.

$$\lim_{x\to\infty} \frac{\log x}{x^n} = \lim_{x\to\infty} \frac{1/x}{nx^{n-1}} = \lim_{x\to\infty} \frac{1}{nx^n} = 0$$

Most of the other indeterminate forms can be reduced to

the two discussed. E.g. if

$$\phi(a) = 0 \qquad \psi(x) \to \infty \quad \text{as } x \to \infty$$

$$\phi(x) \times \psi(x) = \phi(x) \div \frac{1}{\psi(x)}$$

which is of the form 0/0.

The forms 0^0, 1^∞, ∞^0 are reduced by taking logarithms. The form $\infty - \infty$ is usually treated by expanding in series.

Approximate Methods of Integration

The methods of evaluating definite integrals so far given depend on our knowing the indefinite integral explicitly. In practical applications this is frequently not possible. On the other hand, we may have a series of observations from which we can plot a graph. Consider Figure 22.

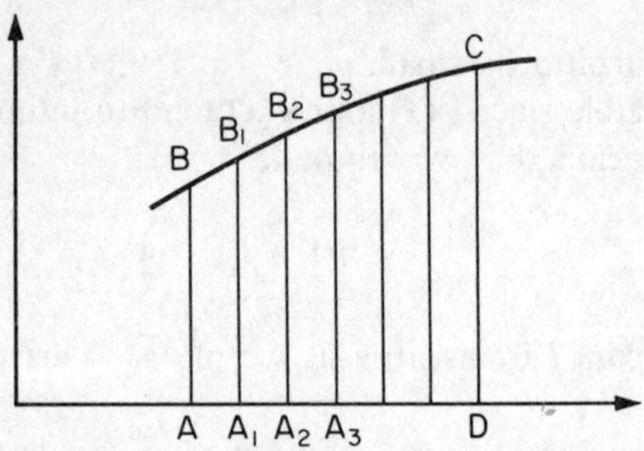

FIG. 22

Suppose we wish to find the value of the area $ABCD$. Let AD be divided into a number of *n equal* parts by $n+1$ vertical ordinates—AB, A_1B_1, A_2B_2, . . ., etc.

A crude approximation to the area can be obtained very readily by supposing that we join the points BB_1, B_1B_2, etc., by straight lines and calculate the sum of the trapeziums so formed. This gives one approximation to the area which we shall call P_1.

Let the equal intervals AA_1, A_1A_2, etc., equal h, and let the vertical ordinates be denoted by $y_0, y_1, \ldots y_n$.

Then $$P_1 = h\left(\frac{y_0+y_1}{2}\right)+h\left(\frac{y_1+y_2}{2}\right)+ \ldots h\left(\frac{y_{n-1}+y_n}{2}\right)$$

Where the graph is convex upwards as shown P_1 is less than the true area. If it is concave upwards it will be greater than the true area.

In order to get another approximation to the area, which we shall denote by P_2, let us draw the tangents at the points corresponding to $y_1, y_3, y_5, \ldots y_{n-1}$ to cut the adjacent ordinates on either side.

Then we have $n/2$ trapeziums each of base $2h$ and forming P_2. Their total area is in excess of the true area if the curve is convex upwards and in defect if concave upwards.

$$P_2 = 2h[y_1+y_3+ \ \ldots \ y_{n-1}]$$

The true area will be between P_1 and P_2.

We can improve on either of these methods to find a third approximation which we shall call P_3.

In the case of P_1 and P_2 we have replaced the curve by sections of straight lines. We shall now replace the curve by sections of a parabola.

By the Second Mean Value Theorem

$$y = F(x) = F(c) + (x-c)F'(c)+\tfrac{1}{2}(x-c)^2F''(x_1)$$

We now assume that if $(x-c)$ is small that we may replace $F''(x_1)$ by $F''(c)$. So

$$y = F(c)+(x-c)F'(c)+\tfrac{1}{2}(x-c)^2F''(c)$$

Form an approximation to the area of a double strip centred at c by evaluating

$$\int_{c-h}^{c+h} y\,dx = \int_{-h}^{h} F(x')\,dx' \text{ when } x' = x-c$$

i.e. $$\int_{-h}^{h}\left(F(c)+x'F'(c)+\frac{x'^2}{2}F''(c)\right)dx'$$

$$= 2h\left[F(c)+\frac{h^2}{6}F''(c)\right]$$

Now if we take $F(c) = y_1$, we have

$$y_0 = F(c-h) = F(c)-hF'(c)+\frac{h^2}{2}F''(c)$$

$$y_2 = F(c+h) = F(c)+hF'(c)+\frac{h^2}{2}F''(c)$$

so $$y_0+y_2 = 2y_1+h^2F''(c)$$

Therefore double strip area is

$$2h[y_1+\tfrac{1}{6}(y_0+y_2-2y_1)] = \tfrac{1}{3}h[y_0+y_2+4y_1]$$

Now, by adding up all the double strips we get another approximation, P_3, to the true area.

$$P_3 = \tfrac{1}{3}h[y_0+y_{2n}+2(y_2+y_4+\;\dots\;+y_{2n-2}) \\ +4(y_1+y_3+y_5+\;\dots\;+y_{2n-1})]$$

This formula is known as Simpson's Rule.

Mean Values

We are familiar with the idea of the arithmetical mean of n quantities, $u_1, u_2, \dots u_n$.

This is $$M = \frac{1}{n}\sum_{r=1}^{n} u_r$$

By analogy we can define the mean value of $F(x)$

$$\frac{1}{b-a}\int_a^b F(x)\,dx$$

Example. The mean value of the ordinate of a semicircle is

$$\frac{1}{2a}\int_{-a}^{a}\sqrt{a^2-x^2}\,dx = \frac{\pi}{4}a = 0{\cdot}7854a$$

This amounts to saying that the area of the semicircle is the same as that of the rectangle $ABCD$ (Figure 23). Note that

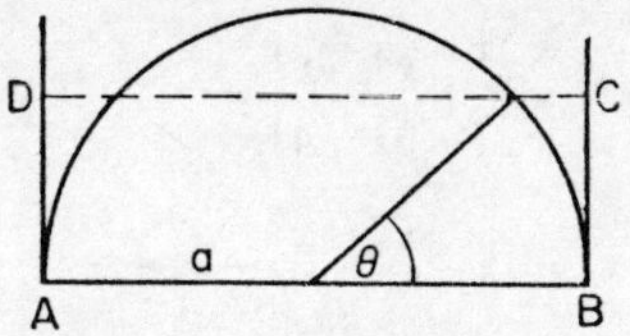

FIG. 23

if we divided the *circumference* into equal intervals and took the mean ordinate we should have got

$$\frac{1}{\pi a}\int_{0}^{\pi} a \sin a \, d\theta = \frac{2a}{\pi} = 0{\cdot}6366a$$

Examples

1. Use Newton's method to find the root of the equation

$$x^3+7x^2-2x-14 = 0$$

which lies between 1 and 2.

2. Using four figure logarithm tables show that the solution of

$$x+\sin x = 1$$

is $x = 29°\ 17'$.

3. Use MacLaurin's form of the remainder in finding R_n for the following functions:

(*a*) $\sin x$ (*b*) $\cos x$ (*c*) e^x
(*d*) $\log(1+x)$ for $0 \leqq x \leqq 1$

Prove that lim R_n is zero in all cases and hence deduce the series for the above functions, viz.

(*a*) $\sin x = x - \frac{x^3}{3!} + \frac{x^5}{5!} - \frac{x^7}{7!} + \dots$

(*b*) $\cos x = 1 - \frac{x^2}{2!} + \frac{x^4}{4!} - \frac{x^6}{6!} + \dots$

(*c*) $e^x = 1 + x + \frac{x^2}{2!} + \frac{x^3}{3!} + \frac{x^4}{4!} + \dots$

(*d*) $\log(1+x) = x - \frac{x^2}{2} + \frac{x^3}{3} - \frac{x^4}{4} + \dots$

4. It may happen that the equation of a curve has two parameters, say α and β, but that these are related. Take the family of straight lines

$$\frac{x}{\alpha} + \frac{y}{\beta} = 1 \qquad \text{(A)}$$

and suppose that there is a relation between α and β

$$\alpha + \beta = a \qquad (a \text{ is constant}) \qquad \text{(B)}$$

We could proceed to find the envelope either by solving B for β and then proceeding as described above, or we may regard β as being known as a function of α and eliminate α, β, and $\partial\beta/\partial\alpha$, i.e. if we have

$$(x, y, \alpha, \beta) = 0$$

with

$$g(\alpha, \beta) = 0$$

and

$$\frac{\partial f}{\partial \alpha} + \frac{\partial f}{\partial \beta}\frac{\partial \beta}{\partial \alpha} = 0$$

with

$$\frac{\partial g}{\partial \alpha} + \frac{\partial g}{\partial \beta}\frac{\partial \beta}{\partial \alpha} = 0$$

These four equations enable us to find the envelope by eliminating

$$\alpha, \quad \beta, \quad \frac{\partial\beta}{\partial\alpha}$$

Now find the envelope above by both methods.

5. Find the envelope of the family of ellipses

$$\frac{x^2}{a^2}+\frac{y^2}{b^2} = 1$$

under the condition that the area ($=\pi ab$) is constant.

6. Prove the following

(a) $\lim\limits_{x\to\infty} \dfrac{\log x}{x^n} = 0$

(b) $\lim\limits_{x\to 0} \dfrac{1-\sqrt{1-x^2}}{x^2} = 0$

7. Show that $\lim\limits_{x\to\infty} (x-\sqrt{x^2-2x}) = 1$

8. Find $\lim\limits_{x\to 0} \dfrac{a^x-b^x}{c^x-g^x}$

9. Taking 10 equal intervals, use Simpson's Rule to show that:

$$\int_0^1 \frac{dx}{1+x^2} = \frac{\pi}{4} = 0{\cdot}78535$$

$$\int_0^{\pi/2} \sin x\, dx = 1{\cdot}00000$$

$$\int_1^2 \frac{dx}{x} = \log_e 2 = 0{\cdot}69315$$

10. The length of a quadrant of an ellipse of semi-major axis l and eccentricity e is given by

$$s = \int_0^{\pi/2} \sqrt{1-e^2 \sin^2 \phi}\, d\phi$$

If $e = \frac{1}{2}\sqrt{2}$, taking six intervals show that $s = 1{\cdot}351$.

11. Find the average value of any point in a circle radius a from the centre.

CHAPTER EIGHT

DIFFERENTIAL EQUATIONS

Definitions

Differential equations fall into two classes—ordinary and partial.

An ordinary differential equation is an equation between a dependent variable y and its derivatives of various orders, and the independent variable x. E.g.

$$\frac{d^2y}{dx^2}+x^2\frac{dy}{dx}+(x^4-x^2)y = \sin x$$

is an Ordinary Differential Equation.

Where we are dealing with functions of two or more independent variables we have equations with partial derivatives. E.g.

$$\frac{\partial^2 z}{\partial x^2}+3xy\frac{\partial z}{\partial y}+\frac{\partial^2 z}{\partial y^2} = e^{-x}\sin y$$

is a Partial Differential Equation.

We shall deal first with ordinary differential equations.

First a few definitions. The *ORDER* of a differential equation is that of the highest derivative which occurs. E.g.

$$\frac{d^3y}{dx^3}+3xy\frac{dy}{dx}+x^2y = 0$$

is an equation of the third order.

The *DEGREE* of a differential equation is that of the highest derivative.

$$x\left(\frac{dy}{dx}\right)^2+\sin x\frac{dy}{dx}+3y = \cos x$$

is an equation of the second degree but the first order.

If the dependent variable y and its derivatives dy/dx, etc., occur only in the first degree the differential equation is a *Linear* Differential Equation. Linear differential equations are of great importance.

The Significance of Differential Equations

As we have seen, a differential equation expresses a relationship between a dependent variable and its various rates of change with respect to an independent variable.

It is a fact that most of what are called Laws of Nature involve rates of change. It is true that some do not. For example, Ohm's Law

$$E = iR$$

which connects the current i to the resistance R in a simple circuit with electromotive force E, does not involve any rate of change.

On the other hand, Newton's Second Law

$$F = m\frac{dv}{dt}$$

relates the impressed force F to the rate of change of momentum of a body of mass m moving in a straight line. We could, taking the x axis as our line, write this

$$m\frac{d^2x}{dt^2} = F$$

To take another example from electrical theory, if we have a capacitor C, the plates of which are joined by a conductor of resistance R and self inductance L, then if V represents the potential difference between the plates as a function of time t, it can be shown that

$$CL\frac{d^2V}{dt^2}+RC\frac{dV}{dt}+V = 0$$

Such equations arise from the mathematical expression of physical laws. We shall now see how differential equations can arise from geometry.

We know from algebra that if we have two equations we can eliminate one quantity, from three equations, two quantities, etc. Now take the equation

$$(x-a)^2+(y-b)^2 = r^2 \tag{8.1}$$

This is the equation of a circle of radius r, whose centre is at the point (a, b). Differentiate with respect to x

$$(x-a)+(y-b)\frac{dy}{dx} = 0 \tag{8.2}$$

Repeating this

$$1+\left(\frac{dy}{ax}\right)^2+(y-b)\frac{d^2y}{dx^2} = 0 \tag{8.3}$$

Now eliminate a and b from Equations (8.1), (8.2), and (8.3). We find

$$r^2\left(\frac{d^2y}{dx^2}\right)^2 = \left\{1+\left(\frac{dy}{dx}\right)^2\right\}^3 \tag{8.4}$$

Now let us examine the differential equations which we have created.

(8.2) is true of circles centre (a, b) no matter what the radius is. It is therefore the differential equation of *all* circles centred at (a, b).

(8.4), on the other hand, is satisfied for all values of a and b if the radius is r. It is thus the differential equation of *all* circles of radius r.

What about (8.3)? Here a and r have disappeared, but b remains, (8.3) is the differential equation of *all* circles whose centres are on the line $y = b$.

Equation (8.2) is a first order equation and if we can by some process 'solve' or 'integrate' it we should arrive back at Equation (8.1). Since, as we shall see, an integral is involved we must expect an arbitrary constant to appear in the solution. This is just the radius 'r' which was eliminated in the process of differentiating (8.1).

So the result of integrating (8.2) is the equation for a circle centre (a, b) and an arbitrary radius. In a practical case we are given information which enables us to determine the arbitrary constants.

For example

$$L\frac{dx}{dt}+Rx = E$$

where L, R, E are constants, has the solution

$$x = \frac{E}{R}+C\,e^{-Rt/L}$$

where C is an arbitrary constant. However, if we are given that at time $t = 0$, $x = 0$, then C must equal $-E/R$ and the required solution is

$$x = \frac{E}{R}(1-e^{-Rt/L})$$

This, in fact, describes how the current x in a circuit of resistance R and self inductance L, and constant e.m.f. E dies away to attain the steady E/R required by Ohm's Law.

We have shown that if we differentiate an equation twice we have three equations and thereby we succeeded in eliminating two constants.

In general by differentiating n times and forming a differential equation of the nth order we could eliminate n constants. We say in general because it does not follow that we can always do this. It may be that less than $(n+1)$ equations will suffice. A rigorous discussion would take us beyond the scope of this book. A converse of the above statement is that the most general solution of a differential equation of order n will contain n arbitrary constants. A rigorous proof is beyond this book.

The solution of a differential equation with the full number of arbitrary constants is called the *Complete Primitive.*

When we have information which enables us to assign definite values to the arbitrary constants we have what is called a *Particular Integral.*

Equations of the First Order and First Degree

Type I. Variables Separable

In this case we can transform the equation to the form

$$f(x)+F(y)\frac{dy}{dx} = 0 \tag{8.5}$$

Where $f(x)$ contains x only and $F(y)$ y only we can, using the notation of differentials, write (7.5)

$$f(x)\,dx+F(y)\,dy = 0$$

$$\therefore \qquad \int f(x)\,dx+\int F(y)\,dy = C \tag{8.6}$$

When C is an arbitrary constant (8.6) is the solution.

Note that a differential equation is said to be solved if we have reduced it to a problem in integration.

Example. $$x^2+2y\frac{dy}{dx} = 0$$

So $x^2\,dx+2y\,dy = 0$. Now integrate and we have

$$\frac{x^3}{3}+y^2 = C \quad \text{where } C \text{ is constant}$$

Type II. Homogeneous Equations

An equation of the first order and first degree is said to be homogeneous if it can be expressed in the form

$$\frac{dy}{dx} = f\left(\frac{y}{x}\right) \tag{8.7}$$

To test this, put $y = vx$ on the right-hand side and we should have an expression in v only.

We use this substitution to solve the equation

Let $y = vx$, then $\quad \dfrac{dy}{dx} = x\dfrac{dv}{dx}+v \quad$ and (8.7) becomes

$$x\frac{dv}{dx}+v = f(v)$$

This may be written

$$\frac{dx}{x} = \frac{dv}{f(v)-v}$$

and so is now separable as type I.

Example. $$2xy\frac{dy}{dx} = x^2+y^2$$

Put $$y = vx$$

$$2v\left(v+x\frac{dv}{dx}\right) = 1+v^2$$

i.e. $$\frac{dx}{x}-\frac{2v\,dv}{1-v^2} = 0$$

$\therefore \quad \log\{x(1-v^2)\} = \log C \quad$ where C is constant,

i.e. $$x^2-y^2 = Cx$$

Type IIa. Equations which can be reduced to be homogenous

$$\frac{dy}{dx} = \frac{ax+by+c}{a'x+b'y+c'}$$

is not a homogeneous type. It can, however, be made so by a change of variable.

The equations $$ax+by+c = 0$$
$$a'x+b'y+c' = 0$$

represent two straight lines which intersect at the point

$$\alpha = \frac{bc'-b'c}{ab'-a'b} \quad \text{and} \quad \beta = \frac{ac'-a'c}{ab'-a'b}$$

Transfer the origin to this point by the substitution

$$x = X+\alpha \qquad y = Y+\beta$$

and the equation reduces to

$$\frac{dY}{dX} = \frac{aX+bY}{a'X+b'Y} \quad \text{which is homogeneous}$$

This fails if the lines are parallel. However the equation is now of the form

$$\frac{dy}{dx} = \frac{ax+by+c}{\lambda(ax+by)+c'}$$

So if we put

$$ax+by = z$$

$$a+b\frac{dy}{ax} = \frac{dz}{dx}$$

and

$$\frac{dz}{dx} = \frac{b(z+c)}{\lambda z+c'}+a$$

which is separable, so of type I.

Type III. Linear Equations

A linear equation is one in which y and dy/dx occur only in the first degree. E.g.

$$\frac{dy}{dx}+Py = Q$$

is a linear equation.

P and Q are functions of x and not of y.

Let us multiply both sides by R where R is a function of x.

$$R\frac{dy}{dx}+PRy = RQ$$

Now the derivative of Ry is, of course, $R\dfrac{dy}{dx}+y\dfrac{dR}{dx}$

and if we can find a function R so that $\dfrac{dR}{dx} = PR$

then the equation takes the form

$$\frac{d}{dx}(Ry) = RQ$$

and this is separable.

We have $$\frac{dR}{dx} = PR$$

and so $$R = e^{\int P\,dx}$$

R is called an *integrating factor*.

Example. $$\frac{dy}{dx}+\frac{y}{x} = x^2$$

$$P = \frac{1}{x} \quad \text{and so} \quad R = e^{\log x} = x$$

So we write

$$x\frac{dy}{dx}+y = x^3$$

$$\frac{d}{dx}(xy) = x^3$$

$$\therefore \quad xy = \tfrac{1}{4}x^4+c$$

Equations Reducible to Linear Type III

$$\frac{dy}{dx}+Py = Qy^n \tag{8.8}$$

Here y occurs on the right-hand side.

But make the substitution $v = y^{-n+1}$

$$\frac{dv}{dx} = (-n+1)y^{-n}\frac{dy}{dx}$$

and we find (8.8) becomes

$$\frac{1}{(1-n)}\frac{dv}{dx}+Pv = Q$$

which is type III.

Occasionally an equation, not linear as it stands, may turn out to be linear if we take y as the independent variable. To test this multiply throughout by dx/dy.

Example. $$(ax+by^2)\frac{dy}{dx}+y = 0$$

this becomes

$$ax+by^2+y\frac{dx}{dy} = 0$$

or $$\frac{dx}{dy}+a\frac{x}{y} = -by^2$$

which is linear.

Type IV. Exact Equations

Suppose we have an equation of the type

$$M+N\frac{dy}{dx} = 0$$

M and N are functions of x and y.

$$M\,dx+N\,dy = 0 \tag{8.9}$$

It may turn out that

$M\,dx+N\,dy$ is a complete differential, i.e. there exists a function $u=u(x, y)$ such that

$$du = M\,dx+N\,dy.$$

This means, of course, that M and N are functions of a special kind. In fact

$$M \equiv \frac{\partial u}{\partial x} \quad \text{and} \quad N \equiv \frac{\partial u}{\partial y}$$

If this is so then an integral of (8.9) is clearly

$$u = \text{const.}$$

and the equation $M\,dx + N\,dy = 0$ is called an *Exact Equation.* The quick test is whether

$$\frac{\partial M}{\partial y} = \frac{\partial N}{\partial x}$$

Even if this is not so it may be possible to find an integrating factor which makes the left-hand side of (8.9) a complete differential.

Example. $2 \sin y\, dx + \{2 \sin y + (x+y) \cos y\}\, dy = 0$

$(x+y)$ is an integrating factor and the solution is

$$(x+y)^2 \sin y = C$$

Equations of the First Order and Higher Degree

For brevity write p for dy/dx

We deal here only with equations of three types.

(1) Equations which we can solve for p
(2) Equations which we can solve for y
(3) Equations which we can solve for x.

Case (1) Here we can obtain a number of solutions of the type $p = p_r$ where p_r is a function of x and y .On integrating these equations we have in all cases a relation which is a solution of the differential equation.

Example. $$p^2 + p - 6 = 0$$

This gives $$(p-2)(p+3) = 0$$

so $$p = 2 \quad \text{or} \quad p = -3$$

The respective solutions are

$$y = 2x+C_1, \quad \text{and} \quad y = -3x+C_2$$

At first sight, although we had reason to expect two alternative solutions we appear to have two arbitrary constants C_1 and C_2, whereas an equation of the first order should only have one arbitrary constant.

This difficulty is only apparent

$$y = 2x+C_1$$

means all lines parallel to the line $y = 2x$. There are an infinity of values of C_1 corresponding to the infinity of lines.

$$y = -3x+C_2$$

means all lines parallel to $y = -3x$ and here also C_2 can take an infinity of values.

Thus if we replace C_1 and C_2 by C it makes no difference to the solution. We say therefore that we have the alternative solutions

$$y = 2x+C \quad \text{or} \quad y = -3x+C$$

with one arbitrary constant.

Case (2). *Equations solvable for y*

Suppose we have an equation

$$3p^4-py+1 = 0$$

then $$y = 3p^3+\frac{1}{p} \quad \text{differentiating}$$

$$\therefore \quad p = 9p^2\frac{dp}{dx}-\frac{1}{p^2}\frac{dp}{dx}$$

$$\therefore \quad dp\,\frac{(9p^2-p^{-2})}{p} = dx$$

or $$x = \tfrac{9}{2}p^2+\tfrac{1}{2}p^{-2}+C$$

but $$y = 3p^3+\frac{1}{p}$$

If we eliminate p between these equations we have a relation

$$f(x, y, C) = 0$$

and this is a solution.

Case (3). *Equations solvable for x*

Consider $$x = 4p+p^3$$

differentiate with respect to y.

$$\frac{dx}{dy} = 4\frac{dp}{dy}+3p^2\frac{dp}{dy}$$

i.e. $$\frac{1}{p} = \frac{dp}{dy}(4+3p^2)$$

$$\therefore \quad \frac{dy}{dp} = p(4+3p^2)$$

$$\therefore \quad y = 2p^2+\tfrac{3}{4}p^4+C$$

By elimination we can find a solution

$$f(x, y, C) = 0$$

Singular Solutions

We have seen how the solution of a differential equation of the first order is of the form

$$f(x, y, c) = 0 \quad \text{where } c \text{ is a constant.} \qquad (8.10)$$

By varying c this represents a family of curves.

We have seen that if we form the equation

$$\frac{\partial f}{\partial c} = 0 \tag{8.11}$$

and eliminate c between (8.10) and (8.11) we obtain a relation which defines the envelope of the family of curves. Now at any point (x, y) on the envelope, x, y and dy/dx are the same as for some member of the family. Hence all points on the envelope satisfy the differential equation. The expression for the envelope contains no arbitrary constant. This solution is known as the Singular Solution.

A word of warning is necessary. The result of the elimination of c between (8.10) and (8.11) may occasionally contain factors which are not envelopes. Therefore *always* test the singular solution to see if it satisfies the differential equation. Note that it is necessary to solve the equations and obtain (8.10) before we can proceed.

There is a special case of interest called Clairaut's Equation. This is usually written as

$$y = xp+f(p) \tag{8.12}$$

differentiate and rearrange

$$\{x+f'(p)\}\frac{dp}{dx} = 0 \tag{8.13}$$

Hence either $\dfrac{dp}{dx} = 0 \quad p = c$

or $x+f'(p) = 0$

If we put $p = c$ in (8.12) we get

$$y = cx+f(c) \tag{8.14}$$

This is the complete primitive of (8.12).

But if we eliminate p between (8.12) and (8.13) we get a

relation which satisfies (8.13). This is, of course, the Singular Solution.

Example.
$$y = xp + \frac{a}{p} \tag{8.15}$$

The Complete Primitive is easily found to be

$$y = Cx + \frac{a}{C} \tag{8.16}$$

and the Singular Solution is

$$y^2 = 4ax \tag{8.17}$$

The student will no doubt recall that (8.16) represents a family of straight lines all of which touch the parabola

$$y^2 = 4ax$$

There are, through any point in space, two tangents to the parabola (8.1,). If the point is on the envelope, however, the tangents coincide.

Now consider (8.15). We can write this as

$$p^2x - py + a = 0 \tag{8.18}$$

and consider that for any value of x, y we have a quadratic equation for p. This gives the slopes of the two tangents through (x, y). Now express the condition that the slopes coincide. This just means writing down the condition that (8.18) has equal roots

$$y^2 = 4ax$$

and we have once again arrived at the envelope or Singular Solution. We can generalise this result but we shall not give a proof.

Let
$$f(x, y, p) = 0 \tag{8.19}$$

be a differential equation

Form
$$\frac{\partial j}{\partial p} = 0 \tag{8.20}$$

and eliminate p between these equations.

Equations (8.19) and (8.20) express the condition that (8.19) considered as an equation for p has a repeated root.

The result of the elimination will be the envelope and is thus a Singular Solution of (8.19).

The advantage here is that we do not need to solve (8.19) to find the Singular Solution.

Example. The family of circles

$$(x+c)^2+y^2 = r^2 \tag{8.21}$$

has the differential equation

$$yp = \pm\sqrt{r^2-y^2} \tag{8.22}$$

or
$$y^2p^2+y^2-r^2 = 0$$

The condition for equal roots is

$$y^2(y^2-r^2) = 0$$

There appear here to be two answers

$$y = \pm r \quad \text{and} \quad y = 0$$

Testing, we see that $y = \pm r$ is a solution of (8.22) and $y = 0$ is not. The fact is that $y = \pm r$ is the envelope but $y = 0$ is the locus of circles which touch but are not neighbours of the family. This illustrates the warning already given.

Equations of the Second and Higher Order

Before going on to a general discussion, we shall devote some time to discussing a special case which is of great practical importance. This is the case of

Linear Equations with Constant Coefficients

The general case is an equation of the type

$$a_0 \frac{d^n y}{dx^n} + a_1 \frac{d^{n-1} y}{dx^{n-1}} + \ldots + a_{n-1} \frac{dy}{dx} + a_n y = f(x) \quad (8.23)$$

where $a_0 \ldots a_n$ are constants.

We shall discuss equations of the second order. The extension to the third or higher order follows logically. So consider the equation

$$a_0 \frac{d^2 y}{dx^2} + a_1 \frac{dy}{dx} + a_2 y = f(x) \quad (8.24)$$

Let us first take the special case when $f(x) = 0$.

Then

$$a_0 \frac{d^2 y}{dx^2} + a_1 \frac{dy}{dx} + a_2 y = 0 \quad (8.25)$$

The equation (8.25) is called the *Auxiliary Equation*. Before discussing methods of solving (8.25) we should note that if $y = v(x)$ is a solution and $y = u(x)$ is another solution, then

$$y = u(x) + v(x)$$

is also a solution. This statement is easily proved by direct substitution.

To find a solution of the differential equation

$$a_0 \frac{d^2 y}{dx^2} + a_1 \frac{dy}{dx} + a_2 y = 0 \quad (8.25)$$

try $y = e^{mx}$ and we find that this is a solution provided that

$$a_0 m^2 + a_1 m + a_0 = 0 \quad (8.26)$$

Let the roots of this quadratic equation be m_1 and m_2

then clearly $y = A\, e^{m_1 x}$ is a solution and

$y = B\, e^{m_2 x}$ is also a solution

Hence $y = A\, e^{m_1 x} + B\, e^{m_2 x}$ is a solution

and since this has two arbitrary constants A and B it is the Complete Primitive of (8.25). It is, of course, not the Complete Primitive of (8.24) where we have $f(x)$ on the R.H.S.

When we are considering the more general equation (8.24) we refer to the solution of the simpler form (8.25) as the *Complementary Function* (C.F.) The C.F. has two arbitrary constants.

We shall discuss the solution of equations of the type (8.24) in some detail later on. We shall confine ourselves at this stage to mentioning that the procedure is to find *any* solution of 8.24. This will not involve an arbitrary constant and is called a Particular Integral of (8.24). If we now add the Complementary Function we have the Complete Primitive of (8.24).

To be precise, if w is any solution of 8.24 and if the C.F is $Au + Bv$ then the Complete Primitive is

$$y = Au + Bv + w$$

It will be clear that with second order equations, in finding the C.F. we arrive at a quadratic equation in m. For higher order equations we will have the problem of higher order algebraic equations for m. E.g. third order equations lead us to a cubic equation for m, and so on. This can lead to algebraic difficulties. Fortunately most of the equations of importance in physical problems are second order ones. We shall continue our discussion of them.

Let us write the equation for m, as follows

$$am^2 + bm + c = 0 \qquad (8.27)$$

(1) If $b^2 > 4ac$ the roots of (8.27) are real and distinct. The

solution of the equation (8.25) is then

$$y = Ac^{m_1 x} + Bc^{m_2 x}$$

(2) If $b^2 = 4ac$ the roots are equal and the solution might appear to be

$$y = (A+B)\, e^{m'x}$$

but here we have a difficulty. There is only one arbitrary constant since the sum $A+B$ of the two arbitrary constants A and B is really a single arbitrary constant. We have therefore not yet found the solution. We have found a solution of the form $y = Ke^{m'x}$ with one arbitrary constant. As a trial suppose we take the function

$$y = V\, e^{m'x} \quad \text{where} \quad V = V(x)$$

and see what conditions have to be satisfied by V for this to be a solution.

Since we assume equal roots each being m', the original differential equation must be expressible in the form

$$\frac{d^2y}{dx^2} - 2m'\frac{dy}{dx} + m'^2 y = 0$$

Substitute our tentative expression $y = Ve^{m'x}$ and we get

$$e^{m'x}\left(\frac{d^2V}{dx^2}\right) = 0$$

where $V = A+Bx$ where A and B are arbitrary. So the solution is

$$y = (A+Bx)\, e^{m'x}$$

Note that there are now two arbitrary constants A and B so this is the complete solution. The former solution is included (as it had to be) but we have the additional term $Bx\, e^{m'x}$.

(3) If $b^2 < 4ac$, the roots are complex. Let them be

$$M_1 = \alpha + i\beta \quad \text{and} \quad M_2 = \alpha - i\beta$$

A solution is now

$$y = A\ e^{(\alpha + i\beta)x} + \beta\ e^{(\alpha - i\beta x)}$$

If we use the expression of de Moivre this becomes

$$y = A\ e^{\alpha x}(\cos \beta x + i \sin \beta x) + \beta\ e^{\alpha x}(\cos \beta x - i \sin \beta x)$$

or $$y = (C \cos \beta x + D \sin \beta x)\ e^{\alpha x}$$

where C and D are now arbitrary constants. Alternatively, we could write this as

$$y = E\ e^{\alpha x} \sin (\beta x - \gamma)$$

where now E and γ are the arbitrary constants.

So far we have explained the procedure for finding the C.F. When it comes to finding the Complete Primitive of (8.24) we said that we needed to know *any* solution. The question is, how do we find it? Sometimes a particular integral is obvious by inspection, e.g.

$$\frac{d^2 y}{dx^2} + 2x \frac{dy}{dx} + y = 3x$$

Here $y = x$ is a solution.

It will be obvious that this approach will not get us very far and more general methods have to be devised. Before we give these we must discuss the use of Operators.

Algebra with Operators

So far, in developing the subject we have tried to be precise and to give strict logical proofs of all statements made. The subject of differential equations can be treated in this manner, but naturally it makes it more difficult to grasp, and some proofs become extremely long.

We propose, therefore, to approach the subject more from the point of view of the engineer or physicist. He is more interested in getting the answer than in the logical details of how it was obtained. Of course it must be the *correct* answer but to apply strict mathematical rigour in developing the methods of solution is time consuming, however fascinating and important from the point of view of pure mathematics.

This approach has, of course, pitfalls, to which we have already drawn the student's attention. However, if we always *verify* by direct substitution that the solutions we have obtained by the methods about to be discussed are in fact correct, we are on secure ground.

With this preamble let us discuss the use of Operators.

The student will be quite familiar with the idea of algebraic operations with symbols x, y, z, etc., where x, y, z are assumed to stand for numbers.

In carrying out these operations we have certain rules,

e.g. $x+(y+z) = (x+y)+z$ Associative Law I

$x(y+z) = xy+xz$ Distributive Law II

$xy = yx$ Commutative Law III

Most of the algebraic formulae with which we are familiar are deductions from these laws, e.g.

$$(a+b)^2 = a^2+2ab+b^2$$

or the *Index Law*

$$x^m \times x^n = x^{m+n}$$

provided we have *defined* x^m.

Now there is no particular merit attaching to the above laws. They are merely the rules of the game. All we really require is that they shall not lead us to inconsistency or to results which are contradictory.

Nor need the symbols necessarily refer to *numbers*. If they do not refer to numbers we must, of course, *define* what we mean by the operations of addition and multiplication. For example, let x, y, etc., refer to matrices. Then I is true and II is true but III is not.

There are other instances where the Commutative Law is invalid and out of these a whole subject of non-commutative algebras has been built up.

The Operator *D*

We define the symbol D as follows:

$$Dy \quad \text{means} \quad \frac{dy}{dx}$$

Note that D by itself is not defined and Dy means that we operate on y.

Now $$D(Dy) = \frac{d}{dx}\left(\frac{dy}{dx}\right) = \frac{d^2y}{dx^2}$$

so if we write $D(Dy)$ *as* D^2y, this means $\dfrac{d^2y}{dx^2}$

It begins to look as if we could treat D as an algebraic symbol, but before doing so we must check that it obeys the Laws of Algebra.

Law I is irrelevant since D cannot stand alone.

Law II is obeyed since

$$D(y+z) = Dy+Dz$$

Law III is true for constants since

$$D(cx) = c(Dx)$$

But it is NOT true for variables since

$$D(xy) = x(Dy)+y(Dx)$$

Now consider the equation

$$\frac{dy}{dx}-\alpha y = X \tag{8.28}$$

We already know the particular integral. It is

$$y = e^{\alpha x} \int e^{-\alpha x} X\, dx$$

Now write the symbol D as explained above for the operator d/dx, (8.28) becomes

$$(D-\alpha)y = X$$

or
$$y = \frac{1}{D-\alpha} X \tag{8.29}$$

Can we assign a meaning to (8.29)?

Its known solution suggests that we *can* do this. We agree to interpret

$$\frac{1}{D-\alpha} X = e^{\alpha x} \int e^{-\alpha x} X\, dx$$

In the important special case of $\alpha = 0$ we deduce

$$\frac{1}{D} X = \int X\, dx$$

This now gives us a meaning for D^{-1}.

$D^{-1}X$ is the indefinite integral $\int X\, dx$.

This might have been expected but it required to be demonstrated. These general results may be very much simplified—when the function X has special forms.

1. $$X = e^{ax}$$

$$DX = aX \qquad D^2X = a^2X \quad \text{etc.}$$

so $$a_0 D^n X + a_1 D^{n-1} X + \dots = a_0 a^n X + a_1 a^{n-1} X + \dots$$

This *suggests* that if our differential equation is

$$F(D)y = e^{ax} \quad \text{where } F(D) \text{ is a polynomial in } D$$

So that
$$y = \frac{1}{F(D)} e^{ax}$$

then a Particular Integral will be given by

$$y = \frac{1}{F(a)} e^{ax}$$

provided, of course $F(a) \neq 0$

Example. $(D+5)^2 y = e^{2x}$

A P.I. is $y = \dfrac{e^{2x}}{(D+5)^2} = \dfrac{e^{2x}}{49}$

For the Complete Primitive we must add the C.F. so

$$y = (A+Bx)\, e^{-3x} + \frac{e^{2x}}{49}$$

II. $X = \cos ax$

In this case $F(D^2)X = F(-a^2)\cos ax$

e.g. $(D^2+3D+1)y = \cos 2x$

A P.I. is

$$y = \frac{\cos 2x}{-4+3D+1} = \frac{\cos 2x}{3D-3} = \frac{(3D+3)\cos 2x}{9D^2-9}$$

$$= \frac{(3D+3)}{-45}\cos 2x = \frac{(D+1)\cos 2x}{-9}$$

$$= \tfrac{1}{9}[2 \sin 2x + \cos 2x]$$

III. $X = x^m$

In this case we try to expand $1/(F(D))$ as a polynomial in ascending powers of D using the binomial theorem.

From the definitions it is easy to show that the expansion of $(D+\alpha)^n$ by the binomial theorem is valid if n is an integer and α is a const. We shall assume that the expression for a negative integer is also true. E.g.

$$(D^2+9)y = x^2$$

$$y = \frac{1}{D^2+9}x^2$$

$$= \frac{1}{9}\left(1-\frac{D^2}{9}+\frac{D^4}{81}\ldots\right)x^2 = \frac{1}{9}\left(x^2-\frac{2}{9}\right)$$

Hence the Complete Integral is

$$y = \tfrac{1}{9}(x^2-\tfrac{2}{9})+A\sin 3x+B\cos 3x$$

A theorem which is sometimes useful is the following:

$$F(D)\{e^{ax}V\} = e^{ax}F(D+a)V$$

V is any function of x and F is a polynomial in D.

The proof follows from the derivative of a product.

$$D(e^{ax}V) = e^{ax}DV+a\,e^{ax}V = e^{ax}(D+a)V$$

$$D^2(e^{ax}V) = e^{ax}(D+a)^2V$$

and finally

$$D^n(e^{ax}V) = e^{ax}(D+a)^nV$$

Since $F(D)$ is a polynomial in D the result follows. The case $V = 1$ is occasionally useful.

Example. $(D^2+9)y = \sin 3x$

We cannot apply direct methods here because if $D^2 = -9$ the denominator would vanish in the expression

$$y = \frac{\sin 3x}{D^2+9}$$

But $\sin 3x$ is the imaginary part of e^{3ix}.

Write
$$y = \frac{1}{(D-3i)(D+3i)}e^{3ix}$$

$$= \frac{1}{(D-3i)}\left[\frac{e^{3ix}}{6i}\right]$$

Now use the previous theorem taken $V = 1$.

$$\frac{1}{D-3i}\left[\frac{e^{3ix}}{6i}.1\right] = \frac{e^{3ix}}{6i}\frac{1}{D}.1 = \frac{x\,e^{3ix}}{6i}$$

Since we can interpret $1/D$ as $\int dx$

$$\therefore \quad \text{the P.I. is } -\frac{x \cos 3x}{6}$$

$\therefore$ the solution is

$$y = A \cos 3x + B \sin 3x - \frac{x}{6} \cos 3x$$

Homogeneous Linear Equations

These are of the form

$$(a_0 x^n D^n + a_1 x^{n-1} D^{n-1} + \ldots + a_n)y = f(x)$$

We can make this into an orthodox linear type with constant coefficients by putting

$$x = e^t \quad \text{so} \quad D = \frac{1}{x}\frac{d}{dt}$$

and we have $(a_0 D_1^n + a_1 D_1^{n-1} + \ldots + a_n)t = F(t)$

when D_1 means the operator d/dt.

Simultaneous Linear Equations with Constant Coefficients

This type of problem occurs in the theory of small vibrations about a state of equilibrium in a conservative dynamic space. Here $D \equiv \dfrac{d}{dt}$.

Example.

$$(2D^2 - 4)x - Dy = 2t$$

$$2Dx + (4D-3)y = 0$$

Eliminate y then

$$\{(4D-3)(2D^2-4) + 2D^2\}x = (4D-3)2t$$

$$(8D^3-4D^2-16D+12)x = -6t+8$$

$$4(2D^3-D^2-4D+3)x = 8-6t$$

$$4(D-1)^2(2D-3)x = 8-6t$$

Now eliminate x and we find

$$4(D-1)^2(2D-3)y = -4$$

In this case we can determine the P.I.s by inspection. For x it is $-\frac{1}{2}t$ and for y it is $-\frac{1}{3}$. Hence

$$x = (A+Bt)\, e^t + C\, e^{-3/2t} - \tfrac{1}{2}t$$

$$y = (L+Mt)\, e^t + N\, e^{-3/2t} - \tfrac{1}{3}$$

Now, we have to substitute these solutions into the equations to determine the relation between the constants ABC, LMN.

We shall find that

$$C+3N = 0, \qquad 2B+M = 0, \qquad 2A+2B+L+M = 0$$

This will now give us a solution with three arbitrary constants.

General Methods for Solving Equations of the Second Order

When one integral of the Complimentary Function is known, say $y = z$, we can reduce the problem to that of solving an equation of the first order.

Let the original equation be

$$\frac{d^2y}{dx^2}+P\frac{dy}{dx}+Qy = R \tag{8.30}$$

where P, Q, R are functions of x.

We now assume that $y = z$ is a solution of the C.F. and put $y = Vz$ where V is a new variable.

then $$\frac{dy}{dx} = Z\frac{dV}{dx}+V\frac{dz}{dx}$$

and $$\frac{d^2y}{dx^2} = Z\frac{d^2V}{dx^2}+2\frac{dz}{dx}\frac{dV}{dx}+V\frac{d^2z}{dx^2}$$

and (8.30) becomes

$$Z\frac{d^2V}{dx^2}+\frac{dV}{dx}\left(2\frac{dz}{dx}+Pz\right)+V\left(\frac{d^2z}{dx^2}+P\frac{dz}{dx}+Qz\right) = R$$

or since $$\frac{d^2z}{dx^2}+P\frac{dz}{dx}+Qz = 0$$

this reduces to

$$z\frac{d(V')}{dx}+V'\left(2\frac{dz}{dx}+Pz\right) = R$$

and this is linear in V' or dV/dx and so the equation is reduced to a simpler type.

The Method of Variation of Parameters

Let our second order differential equation be

$$\frac{d^2y}{dx^2}+P\frac{dy}{dx}+Qy = R \tag{8.31}$$

P, Q, R are functions of x. Now let us suppose that the C.F. of (8.31) is

$$y = \lambda u(x)+\mu v(x)$$

where λ and μ are consts. Now let us try tentatively as a particular integral

$$y = Lu(x)+Mv(x) \tag{8.32}$$

where L and M are functions of x, then

$$\frac{dy}{dx} = L\frac{du}{dx}+M\frac{dv}{dx}+u\frac{dL}{dx}+v\frac{dM}{dx}$$

Let us now impose a condition on L and M which so far

are only related by (8.32). Take this as

$$u\frac{dL}{dx}+v\frac{dM}{dx} = 0 \tag{8.33}$$

then

$$\frac{dy}{dx} = L\frac{du}{dx}+M\frac{dv}{dx} \tag{8.34}$$

hence

$$\frac{d^2y}{dx^2} = L\frac{d^2u}{dx^2}+M\frac{d^2v}{dx^2}+\frac{dL}{dx}\frac{du}{dx}+\frac{dM}{dx}\frac{dv}{ax} \tag{8.35}$$

Substitute in (8.31).

The coefficient of L is $\dfrac{d^2u}{dx^2}+P\dfrac{du}{dx}+Q$ which is zero since u is a solution of the C.F.

Similarly the coefficient of M vanishes and we get

$$\frac{dL}{dx}\frac{du}{dx}+\frac{dM}{dx}\frac{dv}{dx} = R \tag{8.36}$$

If we now solve (8.33) and (8.36) for dL/dx and dM/dx we can integrate the resulting equations and find that

$$L = F(x)+A$$
$$M = G(x)+B$$

where A and B are arbitrary consts. Hence the final solution is

$$y = uF(x)+vG(x)+Au+Bv$$

This method is capable of extension to equations of higher order. The difficulty is that it requires knowledge of the C.F. as a whole and not merely in part as in the previous method.

Examples

Solve the following differential equations:

1. $\dfrac{dy}{dx} = 3xy$

2. $\tan x\dfrac{dy}{dx}+\tan y = 0$

3. $\dfrac{dy}{dx}+(y^2+1) = 0$

4. $(x+2y)\dfrac{dy}{dx} = 2x-y$

5. $\dfrac{dy}{dx} = \dfrac{x-4y+2}{2x-8y}$

6. Solve $\dfrac{d^2y}{dx^2}+\dfrac{3dy}{dx}-4 = 0$

7. $\dfrac{d^2y}{dx^2}+\dfrac{3dy}{dx}-10 = 0$

8. $\dfrac{d^3y}{dx^3}-27y = 0$

9. **Find the singular solution of $y = px + p^2$**

CHAPTER NINE

PARTIAL DIFFERENTIAL EQUATIONS

Definitions

In two dimensions a Partial Differential Equation is a relation between a function z of the independent variables x and y and the partial derivatives of various orders.

E.g. $$\frac{\partial^2 z}{\partial x^2} = x^2 y^2 \frac{\partial^2 z}{\partial y^2} + \sin(x^2+y^2)\frac{\partial z}{\partial x}$$

is a partial differential equation.

The problems associated with P.D.E's. are much more complicated and difficult than those of ordinary differential equations. They are, however, of great importance in physics, and the brief treatment given here will be with this aspect in mind.

Many writers discuss partial differential equations from the point of view of geometry. This has a certain elegance and has other advantages. However, it presupposes a knowledge of differential geometry and does not have the merit of indicating the importance of a given equation in some practical application.

We shall now proceed to develop the subject in an elementary manner with only practical applications in mind.

Physical Basis of Partial Differential Equations

In order to illustrate the physical principles which lie behind a partial differential equation we shall discuss a very simple one—that governing the motion of a flexible string. We shall take as an example a stretched piano string which is a uniform wire stretched between two points. The string is under tension but we shall assume that the tension T is uniform and that the density is also uniform and equal to

ρ per unit length. In practice such strings have an inherent stiffness, but experiments show that the effect of this may be safely neglected. We shall also ignore the effect of air resistance, and also the effects of the weight of the string.

Let the string be held at the points O and L, as represented by Figure 24. Then with the usual notation we can represent the displacement from the equilibrium position, where the string lies straight between O and L, by y where $y = y(x)$.

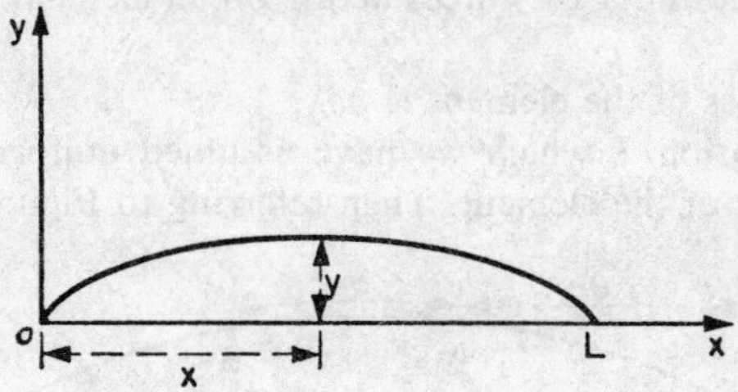

FIG. 24

Since the shape of the string may alter with time and since this means that the displacement y at a point x will vary with time we may express the relationship completely by writing

$$y = y(x, t)$$

i.e. the displacement of any point on the string depends on the time and on its position along the string. y is in fact a function of two *independent* variables x and t. We can now get a physical notion of partial differential coefficients.

$$\left(\frac{\partial y}{\partial x}\right)_{t=\text{const.}}$$

means the *slope* of the profile of the string at the point x.

$$\left(\frac{\partial y}{\partial t}\right)_{x=\text{const.}}$$

means the vertical velocity of the element of the string at the point x.

Let us now try to set up an equation that describes the behaviour of the string.

It is clear that in equilibrium the string is straight and will be along the x axis. When displaced the tensions tend to cause the string to return to this position. There will be equilibrium only when *every* portion of the string is flat.

Let us consider the forces acting on an element δx of the string

The mass of the element is $\rho\delta x$.

The tension T which we have assumed uniform acts at either end of the element. Then referring to Figure 25.

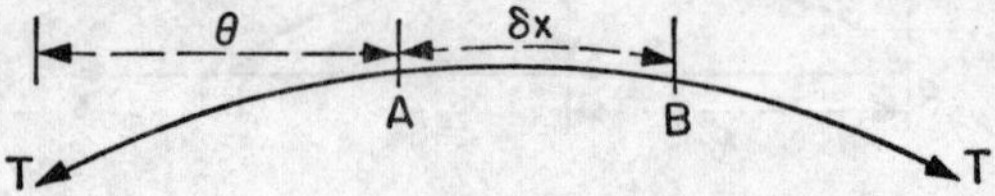

FIG. 25

The vertical component of the tension is

$$-T \sin \theta \quad \text{at} \quad A \quad \text{and} + T \sin \theta \quad \text{at} \quad B.$$

Now θ is a function of x and ϕ is merely the value of θ at the point $x+\delta x$. If the displacement y at any point is small, then so is θ and we may replace the force at A by the expression $T.\dfrac{\partial y}{\partial x}$ since $\dfrac{\partial y}{\partial x} = \tan \theta$ and $\sin \theta \doteqdot \tan \theta$ if θ is small.

The force at B is $T \times \dfrac{\partial y}{\partial x}$ (evaluated at $x+\delta x$).

The Resultant Force on the segment is therefore:

$$T.\frac{\partial}{\partial x} y(x+\delta x) - T.\frac{\partial}{\partial x} y\,(x).$$

Compare this with the remembered definition of a derivative

$$f(x+\delta x)-f(x)$$

and we see that the nett force is

$$T\frac{\partial^2 y}{\partial x^2}\times\delta x$$

Apply Newton's Law to the mass $\rho\delta x$ whose acceleration is $\dfrac{\partial^2 y}{\partial t^2}$ and we get our equation of motion

$$\rho\delta x\frac{\partial^2 y}{\partial t^2} = T\frac{\partial^2 y}{\partial x^2}\delta x$$

or

$$\frac{\partial^2 y}{\partial x^2} = \frac{1}{c^2}\frac{\partial^2 y}{\partial t^2} \qquad (9.1)$$

where

$$c^2 = \frac{T}{\rho}$$

This partial differential equation is a very important one. It is a specially simple case of the Wave Equation.

To see why this name is given, consider the function

$$y = f(x-ct) \quad \text{where } f \text{ is any function.}$$

It is easy to show that this is a solution of (9.1) no matter what f may be. Similarly

$$y = F(x+ct)$$

is a solution of (9.1) no matter what F may be. $y = f(x-ct)$ is a way of stating that if y has a value for any value of x and t, e.g.

$$y_0 = f(x_0-ct_0)$$

if we increase x by the amount ct then the value of y is unchanged.

Now this is what happens when a wave advances with

velocity c (see Figure 26). If we think of a wave on the ocean, the displacement of the water from its level when calm constitutes the wave. This advances with a known velocity, although the water as a whole does not.

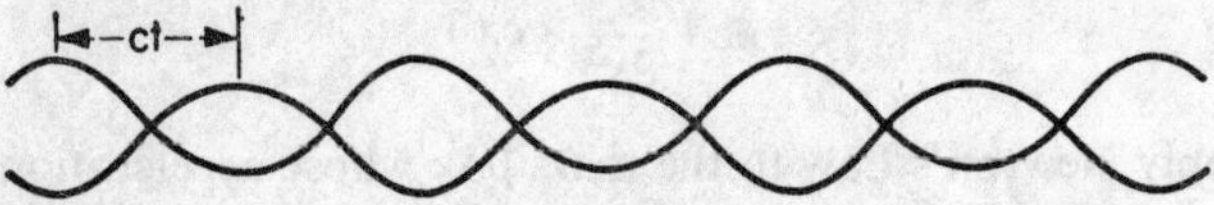

FIG. 26

Although we have found a solution of Equation (9.1), it is an extremely general one and does not tell us as much as we would like.

We derived the equation by considering the behaviour of an element of the string at a point x. The element, so to speak, does not 'know' anything about the rest of the string except that tension is transmitted by neighbouring elements. It has no 'knowledge', for example, that the string is fixed at the two ends.

The solution which we have found reflects this vagueness. It shows us, however, how to calculate the velocity c of the wave motion and shows how it depends on T, the tension, and ρ, the density, of the string.

It is worth mentioning here that by deriving equations for an electro-magnetic field, Maxwell obtained a similar, albeit more complicated, equation for our Equation (9.1), and was thus able to show that electro-magnetic fields could be propagated in space as waves with the velocity of light. This discovery, obtained entirely by mathematical reasoning, was the forerunner of wireless telegraphy and all that has stemmed from it.

Reverting to our string, in any practical problem there are conditions which must be satisfied. For example, our string is not infinitely long, but was fixed at the ends. We can express this by the condition

$$y = 0 \quad \text{at} \quad x = 0$$
$$\text{and} \quad x = l \quad \text{for all values of } t.$$

This is a simple example of a *Boundary Condition*. We must, to be precise, also state something about the condition of the string at some point of time which we take as $t = 0$. For example we could have the condition

$$y = f(x) \quad \text{at} \quad t = 0.$$

This prescribes the shape of the string at time $t = 0$ and is called an *Initial Condition*. Note that the function $f(x)$ must be such that

$$f(0) = 0 = f(l).$$

as given in the boundary condition.

The problem could also be defined by giving the velocity of each element of the string at $t = 0$.

The problem in solving Equation (9.1) is to find a solution which satisfies *both* the initial and the boundary conditions. Let us try to particularise the solution of (9.1). In a real life problem we frequently have a general idea of the type of solution we are hoping to find and to save time we make use of this.

In the case of a stretched string we know that the result will be a vibration. This suggests that $y(x, t)$ will be a periodic function of t.

Let us, therefore, try to see if we can find a solution of the form

$$y = e^{int} \times f(x)$$

We have
$$\frac{\partial^2 y}{\partial x^2} = \frac{1}{c^2}\frac{\partial^2 y}{\partial t^2}$$

$$\therefore \quad \frac{d^2 f}{dx^2} = -\frac{n^2}{c^2} f \tag{9.2}$$

For brevity put $$\frac{n^2}{c^2} = p^2$$

where we use ordinary differential symbols because f is a function of x only.

The solution of the ordinary differential equation (9.2) is known to be

$$f = A \sin px + B \cos px$$

where A and B are arbitrary. A solution of the equation (9.1) is therefore

$$y = (A \sin px + B \cos px)e^{int}$$

We are seeking a solution which satisfies the boundary condition

$$y = 0 \quad \text{when} \quad x = 0 \quad \text{for all values of } t \qquad (9.3)$$

and $$y = 0 \quad \text{when} \quad x = l \quad \text{for all values of } t \qquad (9.4)$$

If we take $B = 0$ we satisfy (9.3) and the condition (9.4) gives

$$\sin pl = 0$$

i.e. $$pl = \pi, \quad 2\pi, \quad 3\pi \ldots$$

or $pl = m\pi$ where m is an integer

We have still to satisfy the initial conditions, i.e. those which must be satisfied for $t = 0$.

Suppose the shape of the string at $t = 0$ is

$$y = g(x)$$

Then $$A \sin \frac{m\pi x}{l} = g(x)$$

for all x where m is an integer. This is obviously impossible. But since the original equation is linear we can add the solutions obtained by taking the various integral values of m, viz. 1, 2, 3, ... and get as a solution

$$y = A_1 \sin\left(\frac{\pi x}{l}\right)e^{\frac{i\pi ct}{l}} + A_2 \sin\left(\frac{2\pi x}{l}\right)e^{\frac{2i\pi ct}{l}} + \dots + A_r \sin\left(\frac{r\pi x}{l}\right)e^{\frac{ri\pi ct}{l}}$$

If we apply our initial condition to this solution we get

$$g(x) = A_1 \sin\frac{\pi x}{l} + A_2 \sin\frac{2\pi x}{l} + \dots A_r \sin\frac{r\pi x}{l} \qquad (9.5)$$

and we ask the question whether we can choose $A \dots A_r$ to satisfy this relationship for all x. Or, to put it in another way, can a function $g(x)$ be expanded as a series of sines of multiples of x?

The answer is that, under certain conditions, it can. We do require an infinite number of terms, i.e. the formula (9.5) is replaced by an infinite series and the various conditions for convergence must be satisfied.

The series $$g(x) = \sum_1^\infty A_r \sin\frac{r\pi x}{l}$$

for $g(x)$ is called the Fourier Series after its discoverer. Let us now return to our solution and write it as

$$y = \sum A_r \sin\frac{r\pi x}{l}\left(ar\cos\frac{r\pi ct}{l} + br\sin\frac{r\pi ct}{l}\right)$$

There are here really only two arbitrary constants for each term which we may write α_r and β_r.

$$y = \sum \sin\frac{r\pi x}{l}\left(\alpha_r\cos\frac{r\pi ct}{l} + \beta_r\sin\frac{r\pi ct}{l}\right)$$

Now the shape of the string at time $t = 0$ gives us a means of finding the values of α_r. They are the appropriate coefficients in the Fourier Series for $y = y(x, 0)$.

We must also, if we are to get a definite solution to our

physical problem, be given the velocity of each point of the string at $t = 0$, i.e. we need to know the value of $\partial y/\partial t$ at $t = 0$. This will give another Fourier Series which gives us the value of the constants β_r.

In an important case where the string is displaced, but released from rest, $\left(\frac{\partial y}{\partial t}\right) = 0$ when $t = 0$, then all β_r are zero.

$$t = 0$$

This leaves the question of finding only the values of α_r. For simplicity let

$$g(x) = \sum_1^\infty \alpha_r \sin rx$$

Multiply by $\sin mx$ and integrate term by term. (Note. This requires justification in any particular case.) We get

$$\int_0^\pi g(x) \sin mx \, dx = \sum_1^\infty \int_0^\pi \alpha_r \sin rx \sin mx \, dx$$

Now $\int_0^\pi \alpha_r \sin rx \sin mx = \frac{1}{2}\alpha_m \pi \qquad \text{if } r = m$

$$= 0 \qquad \text{if } r \neq m$$

$$\therefore \quad \alpha_m = \frac{2}{\pi}\int_0^\pi g(x) \sin mx \, dx \qquad (9.6)$$

Laplace's Equation

This equation, which we have already met in Chapter Six, takes the form

$$\frac{\partial^2 F}{\partial x^2} + \frac{\partial^2 F}{\partial y^2} = 0 \qquad (9.7)$$

or in three dimensions

$$\frac{\partial^2 F}{\partial x^2} + \frac{\partial^2 F}{\partial v^2} + \frac{\partial^2 F}{\partial z^2} = 0 \qquad (9.8)$$

It can, of course, be transferred into polar or cylindrical coordinates and we have seen (p. 77) that for two-dimensional polar coordinates it becomes

$$\frac{\partial^2 F}{\partial r^2}+\frac{1}{r}\frac{\partial F}{\partial r}+\frac{1}{r^2}\frac{\partial^2 F}{\partial \theta^2} = 0 \tag{9.9}$$

All the above equations are linear and lend themselves to the device of assuming a solution which is a product of two functions each containing one and only one of the variables.

Thus in (9.7) we assume $\quad F = X(x)\times Y(y)$

or in (9.9) we assume $\quad F = R(r)\times \Phi(\theta)$

The equations (9.7) and (9.9) now give

$$\frac{\partial^2 X}{\partial x^2} = -\frac{\partial^2 Y}{\partial y^2} \tag{9.10}$$

and

$$\frac{r^2}{R}\frac{\partial^2 R}{\partial r^2}+\frac{r}{R}\frac{\partial R}{\partial r} = -\frac{\partial^2 \Phi}{\partial \theta^2} \tag{9.11}$$

Now in (9.10) the L.H.S. is a function of x only, while the R.H.S. is a function of y only.

The only manner in which they can be equal is for each side to be a constant. We may also replace $\dfrac{\partial}{\partial x}$ by d/dx, etc., since there is only one variable involved in each expression. So let us now put

$$\frac{d^2 X}{dx^2} = -m^2 \quad\text{and}\quad \frac{d^2 Y}{dy^2} = m^2 \tag{9.12}$$

where m is a constant, and similarly for (9.11)

$$\frac{d^2\Phi}{d\theta^2} = -m^2 \quad\text{and}\quad \frac{r^2}{R}\frac{d^2 R}{dr^2}+\frac{r}{R}\frac{dR}{dr} = m^2 \tag{9.13}$$

These are ordinary differential equations and can be solved by the methods of Chapter Eight.

For example, the solution of (9.13) is

$$\Phi = \alpha_m \cos m\theta + \beta_m \sin m\theta$$

where α_m and β_m are constants with

$$R = A_m r^m + B_m r^{-m}$$

when A_m and B_m are constants. Now since the equations are linear, we can add solutions for as many values of the basic constant m as we wish.

So a solution is

$$F = \sum_m (A_m r^m + B_m r^{-m})(\alpha_m \cos m\theta + B_m \sin m\theta) \qquad (9.14)$$

It could happen, of course, that a single term will suffice. In general, however, this solution is appropriate for the problem in which F has a given value on a circle.

Suppose we wish to obtain values of F inside a unit circle and are given that $F = V(\theta)$ at points on the circle. If we assume all $B_m = 0$ then

$$V(\theta) = \sum_m A_m(\alpha_m \cos m\theta) \qquad (9.15)$$

Compare this with the arguments on p. 201, we see that all we need do is to find the Fourier Series for $V(\theta)$ and equate coefficients.

Note that we should strictly *prove* that the sum of an *infinite* number of terms is a solution. This is beyond our scope.

Laplace's Equation is of importance in many electrical problems. It is the equation for the potential in an electrical field over a region in which there are no charges. Physically Laplace's Equation is the expression for the fact that the average value of F over a small sphere is equal to the value of F at the centre, i.e. the field where it applies has no 'lumpiness'. In the two-dimensional case there is an important relationship with functions of a Complex Variable.

If $\quad F(x + iy) = u(x,y) + iv(x,y)$

then $$\frac{\partial u}{\partial x} = \frac{\partial v}{\partial y}, \frac{\partial v}{\partial x} = -\frac{\partial u}{\partial y}$$

So $$\frac{\partial^2 u}{\partial x^2}+\frac{\partial^2 u}{\partial y^2} = 0 \quad \text{and} \quad \frac{\partial^2 v}{\partial x^2}+\frac{\partial^2 v}{\partial y^2} = 0$$

This means then that both u and v are solutions of Laplace's Equation.

Further, the curves given by u = const. and v = const. are orthogonal, i.e. when they intersect it is always at right angles. Hence the families of curves u = const. and v = const. represent the lines of force and the equipotential lines in *some* electrical problems. Sometimes, therefore, we can find solutions of Laplace's Equation by studying functions of the complex variable $Z = x+iy$.

Laplace's Equation is often written for brevity as

$$\nabla^2 F = 0 \tag{9.16}$$

The Equation for the Conduction of Heat

This is the equation which defines the flow of heat by conduction in a solid body. It is also applicable in various diffusion problems.

In a single dimension it is

$$\frac{\partial \theta}{\partial t} = K\frac{\partial^2 \theta}{\partial x^2} \tag{9.17}$$

This gives the value of θ, the temperature in the problem of heat flow in a long bar lying in the x-direction. K is a property of the material, in general a constant called the conductivity.

Example. Show that a solution of (9.17) is

$$\theta = \frac{1}{\sqrt{t}} e^{\frac{-x^2}{4Kt}}$$

This equation can sometimes be solved by the use of operators. In the example given the reader will see for himself how the operational method, so to speak, runs away from us. It develops too rapidly to be followed. Here, as always, we must test the solution. The power of the operational device is well illustrated however.

Write $\qquad D_t$ for $\dfrac{\partial}{\partial t}$ and D_x for $\dfrac{\partial}{\partial x}$

then $$\frac{\partial \theta}{\partial t} = K\frac{\partial^2 \theta}{\partial x^2}$$

becomes $$(D_t - KD_x^2)\theta = 0 \tag{9.18}$$

Thinking of t only and treating KD_x^2 formally as constant for the moment, the solution is

$$\theta = e^{KD_x^2 t} \times A \tag{9.19}$$

A will be a constant as far as t is concerned but may involve x. Expand the R.H.S.

$$\theta = \left(1 + KtD_x^2 + \frac{K^2t^2}{2}D_x^4 + \dots\right)A(x)$$

$$\therefore \quad \theta = A(x) + Kt\frac{d^2A}{dx^2} + \frac{K^2t^2}{2}\frac{d^4A}{dx^4} + \dots$$

Note that this solution gives us θ if we know the value of θ for $t = 0$ to be $\theta = A(x)$.

Heaviside's Equation

As a final example we take the equations for a submarine cable. These are due to Heaviside.

They are

$$-\frac{\partial V}{\partial x} = RI + L\frac{\partial I}{\partial t} \tag{9.20}$$

$$-\frac{\partial I}{\partial x} = KV+C\frac{\partial V}{\partial t} \tag{9.21}$$

Here R is resistance, C capacity, L inductance, and K leakance all measured per unit length and all constants. V and I are the voltage and current and are the variables.

The reader can show as an exercise that a solution is

$$V = V_0 e^{-(a+ib)x+int}$$

$$I = I_0 e^{-(a+ib)x+int}$$

where
$$\alpha^2-b^2 = RK-n^2LC$$

$$2ab = n(RC+LK)$$

and
$$I_0^2(R+iLn) = V_0^2(K+iCn)$$

The interesting point is that the attenuation of the signal along the cable depends on a since the factor e^{-ax} occurs in both the expressions for V and I.

Now a depends on n. This means that the attenuation will be different for different frequencies and that there will be serious distortion of speech, for example, in a long cable.

Heaviside showed that a is independent of n if

$$RC = KL \tag{9.22}$$

It is not possible to arrange this since RC is generally much larger than KL.

However, matters can be improved by making K larger. It is intriguing to ponder whether this solution would ever have occurred to a practical engineer had Heaviside not solved the differential equation!

ANSWERS TO THE EXAMPLES

Chapter Two page 41

1. $20x^3 + 21x^2 + 2$. 2. $3x^2 - 21$. 3. $\dfrac{1 + x^2}{(1 - x^2)^2}$.

4. $\dfrac{x + 1}{2x^{3/2}}$. 5. $\dfrac{1 + \dfrac{1}{2\sqrt{x}}}{2\sqrt{x + \sqrt{x}}}$. 6. $\dfrac{-1}{(1 + x)^2}\sqrt{\dfrac{1 + x}{1 - x}}$.

7. $\dfrac{(3x + 2)^2(6x - 17)}{(2x - 1)^3}$. 8. $\dfrac{dy}{dx} = \dfrac{-(ax + hy)}{hx + by}$.

Chapter Two page 54

1. $m \cos x \sin^{m-1} x$. 2. $n \tan^{n-1} x \sec^2 x$. 3. $-\sin x \cos(\cos x)$.

4. $\tan^3 x[3 \sec x + \cos x]$. 5. $\dfrac{\sin \sqrt{x} + \sqrt{x} \cos \sqrt{x}}{2\sqrt{x}}$

6. $\cot x$. 7. $\dfrac{2x}{\sqrt{1 - x^4}}$. 8. $\dfrac{x}{\sqrt{x^2 - 4}\,(x^2 - 3)}$.

9. $\dfrac{a}{\sqrt{x}(a^2 - x)}$. 10. $e^x(x - 1)^2(x + 2)$.

Chapter Two page 56

1. $\dfrac{-a^2}{(x^2 - a^2)^{3/2}}$. 2. $\dfrac{d^2y}{dx^2} = -4 \sin 2x$, $\dfrac{d^3y}{dx^3} = -8 \cos 2x$.

3. $\dfrac{d^ny}{dx^n} = e^x(x + n)$. 5. $\dfrac{dy}{dx} = \dfrac{1}{a^2 + x^2}$, $\dfrac{d^2y}{dx^2} = \dfrac{-2x}{(a^2 + x^2)^2}$

Chapter Two page 64

3(a). max $y = \frac{1}{2}$ at $x = 1$.
min $y = -\frac{1}{2}$ at $x = -1$.

3(b). Neither max nor min. Inflexion at $x = 0$.

4(a). max $y = \dfrac{1}{e}$ at $x = 1$. 4(b). max $y = A$ at $x = 0$.

5. min $(a + b)^2$.

Chapter Three page 80

1. $e^x \sin y, \quad e^x \cos y.$ 2. $\dfrac{3x^2}{x^3 + y^3}, \quad \dfrac{3y^2}{x^3 + y^3}.$

3. $\dfrac{y}{x^2 + y^2}, \quad \dfrac{-x}{x^2 + y^2}.$

4. $\dfrac{e^{xy}}{x + y}\left[y - \dfrac{1}{x + y}\right], \quad \dfrac{e^{xy}}{x + y}\left[x - \dfrac{1}{x + y}\right].$

5. $\cot (x - y), \quad -\cot (x - y).$

6. $\log \tan (x + y) + \dfrac{x \sec^2 (x + y)}{\tan (x + y)}, \quad \dfrac{x \sec^2 (x + y)}{\tan (x + y)}.$

Chapter Four page 100

1. $-x \cos x + \sin x.$ 2. $-\frac{1}{2}e^{-x^2}.$

3. $\frac{1}{2}x^2 \sin^{-1} x - \frac{1}{4} \sin^{-1} x + \frac{1}{4}x\sqrt{1 - x^2}.$

4. $\frac{1}{2}(x + 1)\sqrt{2x + x^2} - \frac{1}{2} \log (x + 1 + \sqrt{2x + x^2}).$

5. $\frac{1}{2} \sin^{-1} x - \frac{1}{2}x\sqrt{1 - x^2}.$ 6. $\dfrac{1}{4} \log \dfrac{x^2 + 1}{x^2 + 3}.$

7. $2(\sqrt{x} - \tan^{-1} \sqrt{x}).$ 8. $\frac{1}{5}(x^2 - 2)^{5/2} + \frac{2}{3}(x^2 - 2)^{3/2}.$

9. $\dfrac{(2x^2 - 1)\sqrt{1 + x^2}}{3x^3}.$

10. $$\frac{1}{2\sqrt{2}} \log \frac{1 + \tan x - \sqrt{2 \tan x}}{1 + \tan x + \sqrt{2 \tan x}} + \frac{1}{\sqrt{2}} \tan^{-1}(\sqrt{2 \tan x} - 1) + \frac{1}{\sqrt{2}} \tan^{-1}(\sqrt{2 \tan x} - 1).$$

Chapter Five page 119

1. $\frac{1}{5}(b^5 - a^5).$ 2. $\dfrac{1}{2\sqrt{2}} \log \left(\dfrac{\sqrt{2} - 1}{\sqrt{2} + 1}\right).$ 3. $\dfrac{\pi}{8}.$

4. $\dfrac{\pi a^2}{8}.$ 6. $\dfrac{\pi}{6}[(4a^2 + 1)^{3/2} - 1].$

Chapter Six page 145

1. Divergent. 2. Convergent. 3. Convergent if $s > 1$
Divergent if $s = 1$
Divergent if $s < 1$.

4. Convergent.

Chapter Seven page 161

1. 1·412. 4. $\sqrt{x} + \sqrt{y} = \sqrt{a}$.

5. $2xy = \pm \dfrac{A}{\pi}$ where A is area.

8. $\dfrac{\log a - \log b}{\log c - \log g}$. 11. $\dfrac{2a}{3}$.

Chapter Eight page 192

1. $y = Ae^{3x^2/2}$ where A = const. 2. $\sin x \sin y$ = const.

3. $y = -\tan(x + a)$ where a = const. 4. $x^2 - xy - y^2$ = const.

5. $-2x + 4y + 4\log(x - 4y + 4)$ = const. 6. $y = Ae^{-4x} + Be^{x}$.

7. $y = Ae^{-5x} + Be^{2x}$.

8. $y = Ae^{3x} + e^{-3x/2}\left(B\sin\dfrac{3\sqrt{3}}{2}x + C\cos\dfrac{3\sqrt{3}}{2}x\right)$.

9. $y = -\dfrac{x^2}{4}$.

INDEX